The Global Environment

K. Takeuchi M. Yoshino (Eds.)

The Global Environment

With 59 Figures

Springer-Verlag Berlin Heidelberg GmbH

Dr. Kei Takeuchi

Research Center for Advanced Science and Technology, University of Tokyo,
6-1 Komaba 4-Chome, Meguro-ku, Tokyo 153, Japan

Professor Dr. Masatoshi Yoshino

Institute of Geoscience, University of Tsukuba, 1-1-1 Tennodai
Tsukuba City, Tsukuba 305, Japan

Cover photo: Burning Kuwaiti oil well by Alan Harper/The Financial Times

ISBN 978-3-662-01087-7   ISBN 978-3-662-01085-3 (eBook)
DOI 10.1007/978-3-662-01085-3

© Springer-Verlag Berlin Heidelberg 1991

Originally published by Springer-Verlag Berlin Heidelberg New York in 1991

Softcover reprint of the hardcover 1st edition 1991

Typesetting: Camera ready by authors

54/3140-543210 – Printed on acid-free paper

Preface

The IBM Japan International Symposium *Energy and Environment – Global Warming* was held in the Keidanren Guesthouse at the foot of Mt. Fuji, from October 21 to 24, 1990. The symposium was conducted in the context of IBM Japan's longstanding commitment to good corporate citizenship.

On this beautiful planet with its inter-dependent waters, lands and atmosphere, we consider that the problems relating to the global environment are the most serious that the human race will face in the near future.

The symposium provided an opportunity for forty scientists and researchers, from a wide variety of international backgrounds, to address matters relating to the global environment in an international forum. Eighteen papers were presented followed by panel and group discussions, on which the concluding remarks and recommendations are based.

We chose three types of papers to target different aspects of the condition of the global environment: the natural science component; the socio-economic component; and the energy component which links these two.

On the first day the symposium began with a plenary speech by Dr. J. Kondo followed by three keynote speeches, each with a particular focus. The following day, six speakers offered papers relating to the previous day's keynote speeches. On the third morning there was an extensive half-day panel presentation and discussion. Five groups were then formed to consider the following issues: integrating scientific priorities; environment information exchange and dissemination; risk philosophy; technology transfer and efficiency; socio-economic aspects of relations between industrial and developing countries. Concluding remarks were approved after discussion by all participants on the last morning.

In spite of the widely varying academic backgrounds of the participants, discussions were harmonious, productive and insightful. Not only was mutual understanding between people of various nationalities greatly increased, but unanimous agreement was reached on what should be done by the nations of the world, both individually and cooperatively.

On behalf of the organizing committee we would like to acknowledge the cooperation and enthusiasm of all participants, which contributed greatly to the success of the meeting and helped to make it both stimulating and inspiring. We would also like to acknowledge the help of Dr. J. Kondo of the Science Council of Japan and Dr. L. Esaki of the IBM T.J. Watson Research Center who gave us valuable advice and to whom we owe much.

We would like to express our sincere gratitude to Mr. T. Takeda, Mr. S. Ohno and Mr. A. Nemoto of the IBM Japan Secretariat for organizing the conference and arranging the excellent meeting. We would also like to thank Mr. C. McDowell, who helped in the negotiations and arrangements with foreign scholars, and the publisher for the smooth publication of these proceedings. We acknowledge with thanks the work of all the IBM Japan staff who helped to make the conference at beautiful Gotemba, at the foot of Mt. Fuji, a most pleasant occasion.

Additional thanks are due to the authors of the contributions and the panel participants who, although extremely busy, completed their papers promptly to make possible the swift publication of these proceedings of the first in a new series of conferences on environmental issues.

What we can do now is to continue to study the environment in its broader aspects, and we must endeavor to work towards consensus on scientific and humanistic issues with knowledge and wisdom.

We present these proceedings of the symposium on the global environment with the hope that they will help to promote further study in this and related fields, which we believe is of the greatest importance.

Mount Fuji   *K. Takeuchi*
April 1991   *M. Yoshino*

IBM Japan

International Symposium on the Global Environment

Advisory Committee:	J. Kondo	Science Council of Japan
	L. Esaki	IBM T.J. Watson Research Center
Executive Committee:	T. Asai	University of Tokyo
	K. Takeuchi	University of Tokyo
	M. Yoshino	University of Tsukuba
	A. Ichikawa	National Institute for Environmental Studies
Secretariat:	T. Takeda	IBM Japan
	S. Ohno	IBM Japan
	A. Nemoto	IBM Japan

Opening Remarks

Good afternoon, ladies and gentlemen. On behalf of IBM Japan, I would like to say a few words at the opening of the session.

It is a great honor for me to welcome all of you who are taking part in this IBM Japan International Symposium on the Global Environment. I am especially grateful to overseas participants who have come all the way to Japan for this seminar.

Nowadays, it is widely recognized that the issue of the global environment is becoming the most serious problem which human beings face for survival in the future. Of the major environmental issues such as acid rain, ozone depletion and deforestation, I think the problem of global warming is the most critical one. The reason seems to be that this global warming relates directly to the use of energy, which is indispensable to the economic welfare of human beings.

You will notice that these days, if I may use the allegory, the cart, which is called policy has moved far in front of the horse, which is called science. This is, in a way, not an unwelcome step toward better understanding and solution of the problem. However, needless to say, cool and sober scientific analysis remains indispensable. Due consideration should also be given to the cost and benefits of restraints on CO_2 emission.

Having said this, it is clear that global environment problems, and especially global warming, form a new challenge to natural science as well as to social science. This is the reason why we invited distinguished natural as well as social scientists from many countries to participate in this international symposium on the environment with its main focus on global warming.

IBM Japan is continuing its efforts to contribute to society in various ways. The sponsorship of this symposium is one aspect of this "corporate social responsibility". As an organizer of this symposium, I would like to express my deep gratitude to Dr. Kondo, and to the members of the Executive Committee, Prof. Asai, Prof. Ichikawa, Prof. Takeuchi and Prof. Yoshino for their great devotion and wide-ranging contributions. We, the secretariat of the symposium, sincerely hope that lively and fruitful discussions will take place during the symposium, and we trust that all of you will enjoy your stay in this Keidanren Guest House at the foot of Mt. Fuji at the best time of the year in Japan. Thank you.

IBM Japan Ltd.   *T. Takeda*

Contents

Plenary Lecture:
Research and Policy on the Environment in Japan

J. Kondo

Science Council of Japan, 22-34 Roppongi 7-Chome,
Minato-ku, Tokyo 106, Japan

It is indeed a great honor for me to read the Plenary Lecture at the opening of the IBM Japan International Symposium on the Global Environment. My topic is research and policy on the environment in Japan. Now ladies and gentlemen, when you see the globe from space, you can see that our Earth is but a tiny speck in the infinitely large universe. Mr. Yuri Aleckseyevich Gagarin was the first man who looked from space at the Earth and said, "Our Earth is blue". As you see from this, our planet is so small. On this model of the globe, how thick do you imagine the atmosphere to be? This terrestrial globe is about 20 cm in diameter. The Earth is really an ellipsoidal sphere, and 90% of the mass of the atmosphere is concentrated only 20 km from the Earth. If you imagine it on this model of the Earth, the thickness of the atmosphere would be only one-tenth of a millimeter. Therefore, you can understand how easily the infinite-appearing sky, which covers the whole world, is damaged by human activities.

From the middle 1960s to the 1970s, in Japan we had serious problems of regional environmental pollution. However, at the present time, we believe that we have been able to control the pollution of the atmosphere. This kind of pollution happened locally in some places in this country. But now, what we are worrying about are global environmental problems, because the atmospheres over each country are connected. Therefore, we have to consider global environmental problems, about which I would like to talk this afternoon.

The real cause of global environmental problems is the rapid rate of growth of the world's population. As you can see in this logarithmic scale (Fig. 1), it was 1987 when the population exceeded 5 billion. At the end of this century, the total population will surpass 6.3 billion, and this rapid increase of world population may affect the global environment. At the time of the Industrial Revolution, the population was one billion. Now it has reached over 5 billion, and will reach 10 billion by the end of the next century.

Figure 2 shows human energy consumption throughout human history. In prehistoric times people only needed to eat. Per capita, they needed 2,500 kcal/day. Then came the ages of hunting and agriculture - about 5,000 B.C. - and then the development of cities appeared in the Middle East, in Greece, in southern Europe and in China. After the Industrial Revolution, we used much more energy for food, housing and commerce, agriculture and transportation. Now, we are using an amount of energy, which is one hundred times that used in primitive ages. If the population itself has increased, the

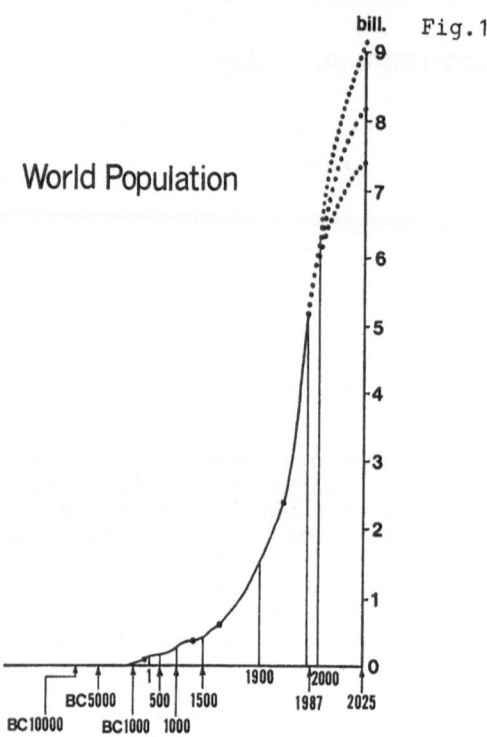

bill. Fig.1

World Population

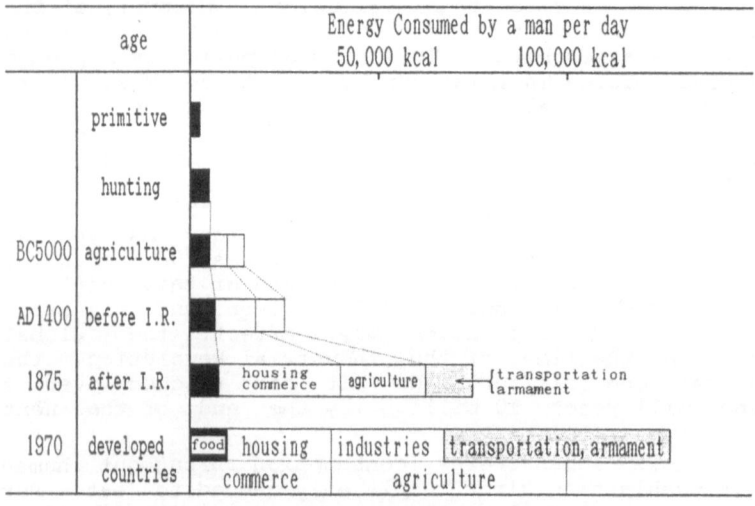

	age	Energy Consumed by a man per day 50,000 kcal 100,000 kcal			
	primitive				
	hunting				
BC5000	agriculture				
AD1400	before I.R.				
1875	after I.R.	housing commerce	agriculture	transportation armament	
1970	developed countries	food	housing commerce	industries agriculture	transportation, armament

Fig.2. Change of energy consumption

consumption of energy has increased one hundred times. Consequently, if the world population is doubled by the end of the next century, the amount of energy consumed in the world will also be doubled. However, this is not true, because the developing countries want to live more comfortably, more conveniently. In summer they need air-conditioning, and in winter they need heating. Therefore, when you consider the situation in the developing countries, if the population of the world doubles, the total consumption of energy will be tripled, or even more - maybe it will reach five times the present value.

With regard to environmental problems, we are experiencing global warming due to the "Greenhouse Effect", an ozone hole caused by CFCs, deforestation aggravated by acid rain, ocean pollution, desertification, and urbanization in developing countries. These are serious problems of the global environment. Because of lack of time, I cannot go into them in detail. Sometimes, through damage to tankers, we have disasters causing oceanic pollution.

We remember that the bay of Alaska was polluted by 40,000 kl oil by the crash of Exxon's tanker in Alaska in March 19, 1990.

Now I would like to concentrate on the problem of global warming. The evidence of CO_2 increasing in the atmosphere is well known. This concentration of CO_2 has been studied by Dr. C. Keeling on Mauna Loa in the Hawaiian Islands. Figure 3 indicates average monthly data. If you look at the annual average, it appears to have accelerated in recent years. This is a big problem, because the concentration of CO_2 in the atmosphere will cause global warming - raising the temperature of the atmosphere. We will refer to this table later on, but I would like to simply indicate some of the recent data with regard to CO_2 emissions.

THE RISE IN ATMOSPHERIC CARBON DIOXIDE

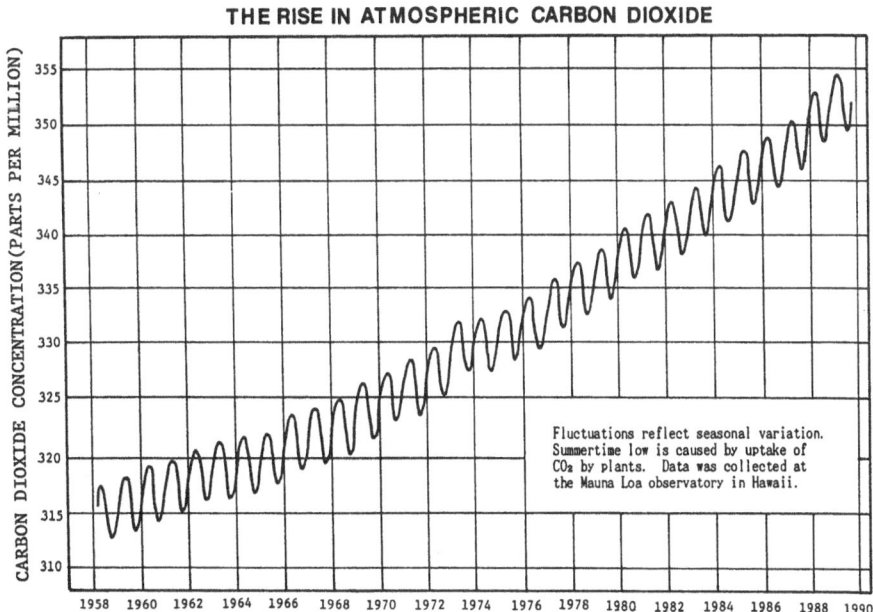

Fluctuations reflect seasonal variation. Summertime low is caused by uptake of CO2 by plants. Data was collected at the Mauna Loa observatory in Hawaii.

Fig.3

Fig.4

We polluters Net per head emissions

of greenhouse gases, 1987 (tonnes of carbon)

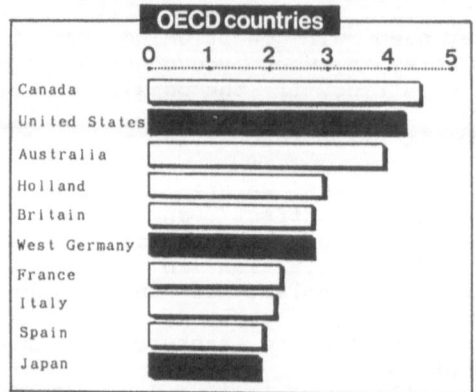

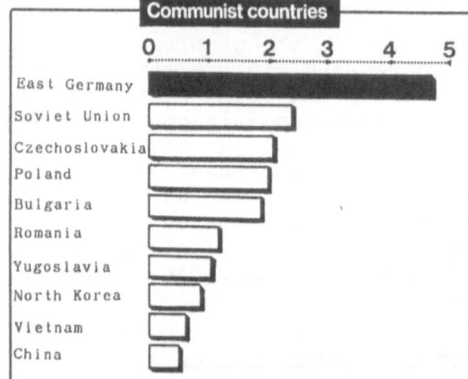

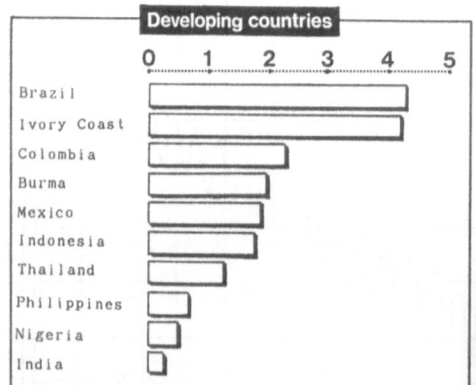

Figure 4 indicates the per capita greenhouse gases (mostly CO_2 emissions) of each country. Notice the OECD countries: the largest polluter is Canada because in Canada, they have large resources of energy, while the population is not so

Per Capita CO₂ emission(1985)

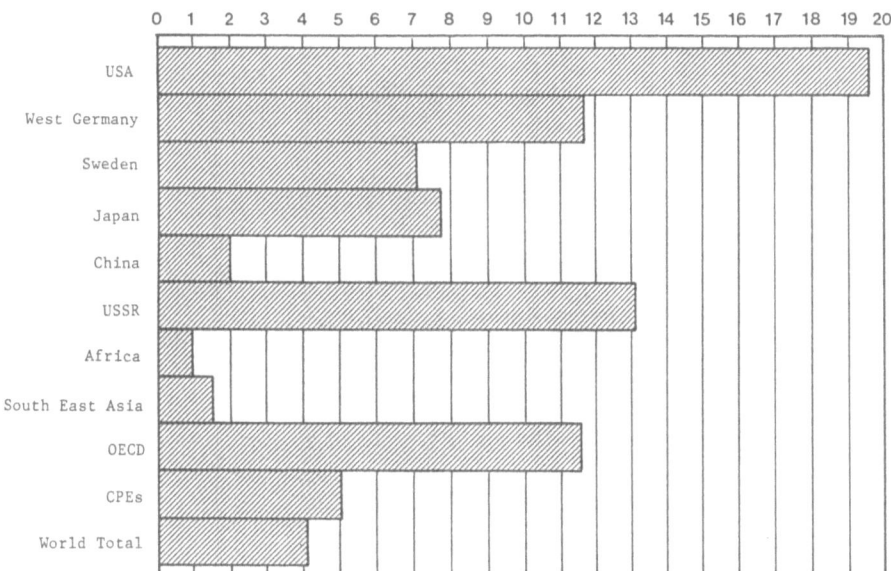

Fig.5

large. Their per capita emission of CO_2 is the highest in the
world. Second is the United States, followed by Australia,
Holland, Britain, West Germany, France, Italy. Last, is Japan
where per capita emission of CO_2 is very small compared with
the other developed countries. Of the communist countries at
that time   this is 1987 data - East Germany has the highest
rate. However, at the beginning of this month (October,
1990), East and West Germany unified. I would like to
congratulate them. But I am sorry to say, Prof. Bach, that
you have to take into account the emissions of East Germany
from now on!
     If you look at the less-developed countries - Brazil, Ivory
Coast, Colombia, Burma, Mexico, and India - they consume very
small amounts of energy, but in total rather large amounts,
because the population is very large, while the per capita
consumption is low.
     These bars (Fig. 5) indicate per capita consumption all
over the world. You can see with Japan, that although it
consumes a large amount of energy, per capita emission rate of
CO_2 is rather low. The USA, the USSR and China, and the
average for all the OECD countries is also shown.
     What is very difficult is that CO_2 is not the only gas
contributing to the global "Greenhouse Effect". There are
many gases which contribute to the rise in temperature in the
atmosphere. Figure 6 indicates the contribution of other
gases: CO_2, methane, NO_x, CFCs, ozone, and others.
     This figure also indicates the warming contribution per
decade up to 2030. We expect that CO_2 itself will be

5

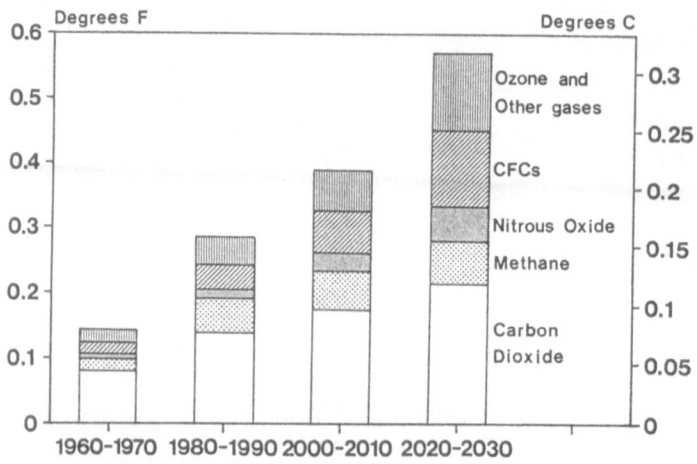

Source: World Resources Institute, based on radiative forcing
estimates in Ramanathan, 1985a. Op. Cit.

Fig.6. Past and future warming contribution for primary greenhouse gases (warming per decade)

increased, as I said before, because of the burning of fossil
fuels - oil and coal. Since we have agreed to stop the
production of CFCs, they should be eliminated after the year
2000. This figure indicates how these "greenhouse" gases have
contributed to global warming in each decade.

Figure 7 looks somewhat complicated; however, it indicates
causes and effects in the changing global environment. The
fundamental cause is, as I said before, world population
increase. Man consumes more food and energy and socio-
economic activities are increased. The emissions of SO_2 and
NO_x are then increased. Also CFC discharge is increased, due
largely to industrial activities. These increases result in
the depletion of the ozone layer, which may cause increased
skin cancer. The increased SO_2 and NO_x cause acid rain, and
because of this, deforestation will take place. The "flag"
symbol indicates the effect of the environment, and the "box
flag" symbol indicates the global effect. If the level of CO_2
increases, temperatures rise - "Greenhouse Effect" - then CO_2
evaporates from the oceans. The lines indicate that
increasing CO_2 has a positive feedback effect: when the
temperature rises, CO_2 from the oceans evaporates, and then,
together with the temperature rise and the evaporation from
the oceans, the CO_2 concentration itself is increased. Thus
we have positive feedback causing increased CO_2 concentration.

If deforestation takes place, CO_2 will increase, because of
decrease of the plant function of photosynthesis (Green trees
absorb CO_2). Rising temperatures will be followed by climate
change, ecology change, and rising sea levels. These are the
main topics, I believe, of this Symposium, at which we have so
many distinguished scientists from many fields. At the
present time in the world, there are many congresses,
symposiums, and panels of this kind taking place. An example
is IPCC. At this particular meeting, however, we would like

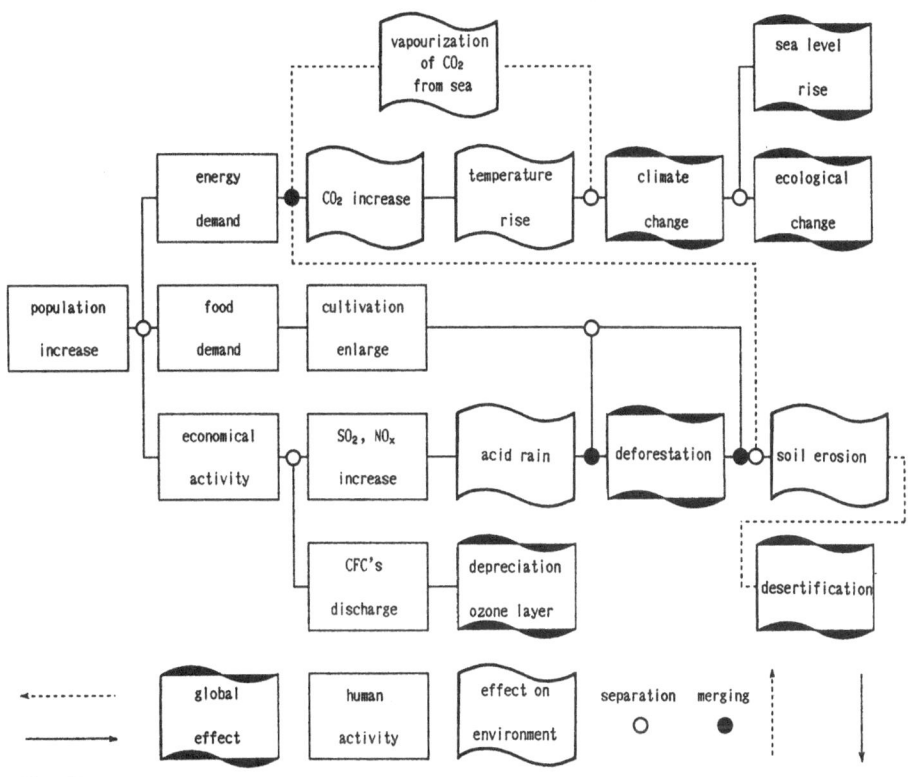

Fig.7

to discuss these problems from economic, social, scientific, and technological standpoints. I am looking forward very much to hearing your views.

When the sea level rises, what will happen? This is an important question. Here I have a picture (Fig. 8) which was drawn by a famous Japanese artist, Hokusai in 1831 about 150 years ago. You can easily get this postcard if you go to a bookstore at Narita Airport, or in one of the hotels. If the sea level rises the ocean will come closer to this meeting place. A large part of the land of Japan will be sunk in the sea. What is still unknown is whether the sea surface will absorb CO_2 or discharge CO_2. Is the ocean the source or the "sink" of CO_2? We still do not know exactly. This is one kind of "uncertainty". From the scientific point of view, we must study the effects of the ocean, the interaction between the ocean and the atmosphere, and the interaction between the ocean and the land.

The Science Council of Japan has proposed to the Japanese government that they should increase their effectiveness by joining IGBP - International Geosphere Biosphere Problem, which is going on through the international cooperation of scientists under the leadership of ICSU - the International Council for Scientific Unions.

Figure 9 indicates how much total CFC gases will increase in different parts of the world. The solid black indicates

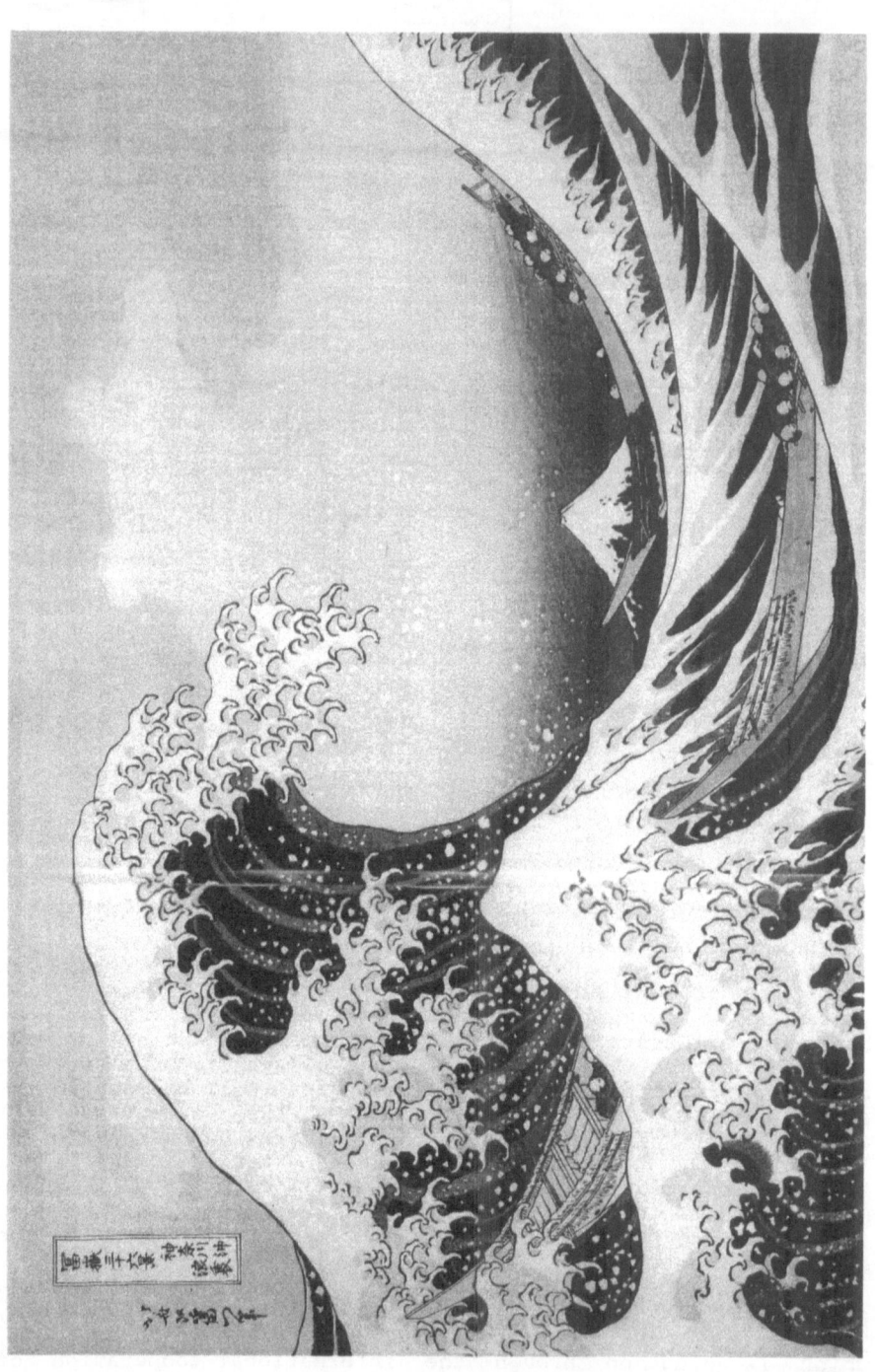

Fig. 8

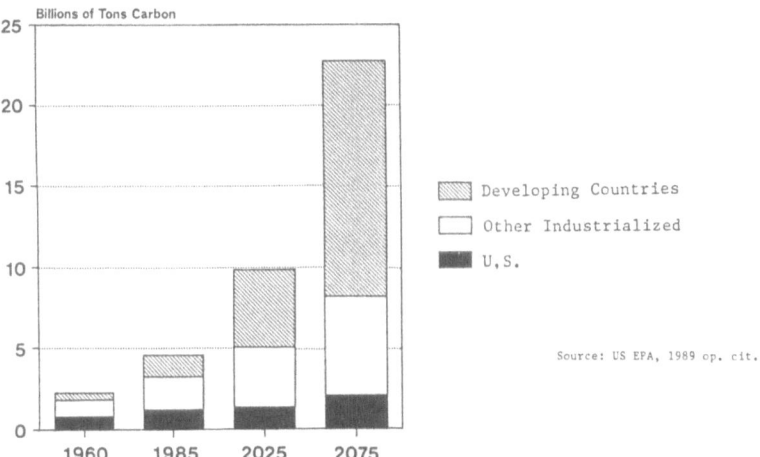

A SCENARIO OF DEVELOPING COUNTRIES AS   Fig.9
PAST AND FUTURE EMITTERS OF CARBON

Developing Countries

Other Industrialized

U.S.

Source: US EPA, 1989 op. cit.

the US, the white indicates other industrialized countries
including Japan. The hatched area develops quickly until the
end of 2075 which indicates increased production of CO_2 due to
economic expansion. It has been announced that Japan will
keep emissions at their present level. This means that by the
21st Century the level of CO_2 emissions should be the same as
in the year of 1990. But how are we to control emissions from
developing countries is a big problem. In China they have
large natural resources of coal, and they could therefore use
coal as the main energy source. In East Germany they have
"brown coal", or lignite, and they burn brown coal as a source
of electricity. However, they do not have devices to absorb
CO_2. We in the developed countries have to do something for
the developing countries.
Many people talk about "sustainable development". The term
came from UNWCED (The World Commission of Environment and
Development). They published the well known report "Our
Common Future" from the Oxford Press in 1987. The term
"sustainable development" first appeared in this report. Many
people are concerned about sustainable development, but few
people discuss how to sustain development in developing
countries. I think that, as a strategy for sustainable
development, the first step is to increase the efficiency of
energy consumption: increase the efficiency of boilers,
generators, and so on. Secondly we have to reconsider
production processes, so as not to consume so much energy.
Thirdly we must look at the construction of anti-pollution
devices. Recycling of materials will also be useful.

What I said previously about increasing efficiency is one
of the key strategies for achieving sustainable development.
Increasing efficiency means greater output with the same
amount of input. The efficiency of the upper case in Figure
10 is much higher than the lower one. If we do not fully
utilize resources, then that part of the resource which is
unused becomes the waste which causes environmental pollution.

9

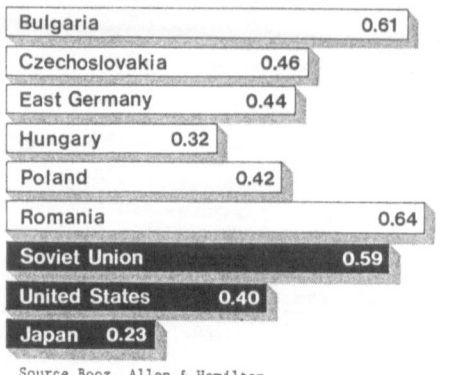

Output | Pollution

Efficiency = Output / Input

Output | Pollution

Fig.10   Fig.11

Increase in efficiency brings about a decrease in pollution.
Recycling is also useful as a means of reducing the unutilized
proportion of the resource.

I took Figure 11 from last month's New York Times. It
shows the number of metric tons of oil - or the energy
equivalent in coal or gas - used to produce $1,000 of goods
and services in selected Eastern European countries. Japan
needs only 0.23 metric tons to produce the same value of
products. One metric ton is equal to 7.3 barrels, so you can
easily calculate the amount of oil consumed. As you can see,
the values of European countries are large, while today
Japan's energy consumption is relatively small. Japan is
using resources very efficiently.

The fourth column of Table A shows CO_2 emissions per GNP.
Sweden comes lowest in this category. This is because they
use large amounts of nuclear energy which does not emit CO_2.
In China because efficiency is very low they emit large
amounts of CO_2.

Something most important is shown in Figure 12. This is
the experience of Japan between 1970 and 1977. You can see
that the concentration of SO_2 was considerably reduced. At
the same time, GNP increased. Energy consumption ran somewhat
parallel to GNP. These changes were because of our experience
of the "Oil Shock". In addition we had bad environmental
pollution and many people were suffering from asthma or lung
diseases. As a result Japanese people took measures to reduce
the pollution of the atmosphere and the consumption of energy.

Figure 13 indicates the extent to which we increased
capacity for desulfuration between 1970 and 1978, in terms of
the number of plants which introduced equipment capable of the
absorption of SO_2. The figure shows that the capacity for
desulfuration in normal cubic meters per hour is increasing,

International Comparison of CO_2 related data(1985)

item country	Per Capita			Per GNP Million $	
	GNP (US$)	energy consumption(TOE)	CO_2 emission (ton)	CO_2 emission (1000ton)	energy consumption (1000ton)
USA	16,760	7.52	19.63	448	1,171
West Germany	10,270	4.38	11.70	426	1,139
Sweden	11,676	4.86	7.09	416	607
Japan	11,014	3.03	7.75	274	703
China	219	0.60	2.03	2,730	9,242
USSR	4,350	4.94	13.11	1,135	3,013
Africa	613	0.35	0.99	573	1,608
South East Asia	842	0.56	1.53	662	1,813
OECD	10,836	4.61	11.55	425	1,065
CPEs	1,610	1.73	5.05	1,073	3,137
World Total	2,681	1.55	4.13	577	1,539

Trend of SO₂ Emission

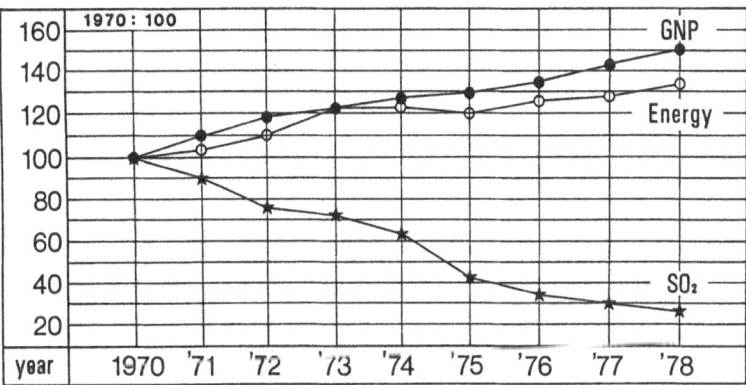

Fig.12

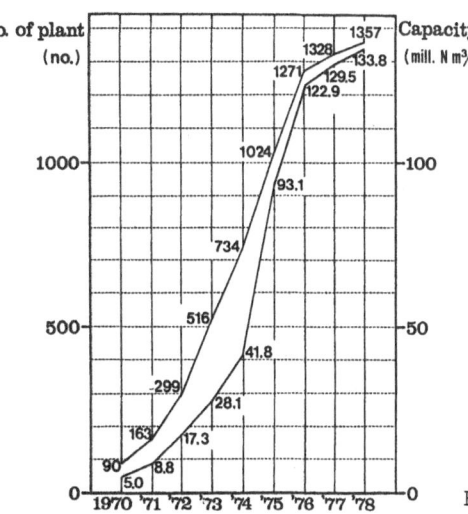

Trend of Desulphication Capacity
and Number of Plants

Fig.13

Fig.14. View of downtown Tokyo covered with photochemical smog, looking westward. Mt. Fuji must be seen behind the tall buildings if the air is clear. Summer 1972. (Courtesy of the "Yomiuri" newspaper.)

showing that Japanese industries are making an effort to introduce SO_2 absorption devices.

Here I have a picture (Fig. 14) which I took from "The Daily Yomiuri", which is a very popular newspaper. It is a view of downtown Tokyo in 1972, looking westward from the middle of the city. There was a layer of smog surrounding tall buildings and streets. We could not see Mt Fuji.

In the summer of last year, 1989, the sky over the metropolitan area looked like this (Fig. 15). I took this picture from "The Asahi", which is also a popular newspaper. In this case Mt Fuji can be seen behind the cumulonimbus. In the summer of 1972 the sky was very dirty, but after 16 years it was much cleaner. These two pictures indicate the large effort we have put into absorbing SO_2 emissions.

Although there is some evidence of acid rain this place, the Keidanren Guest House, is surrounded by tall trees, pine trees, birch and maples. It is just the beginning of the beautiful fall in Japan, and I hope that you will enjoy your stay here. If you find trees which are damaged by acid rain, please let me know. Japanese people are very concerned about the effect of acid rain. However, we still do not have forests which are damaged like Schwarzwald (Black Forest) in West Germany, where almost half of the trees have been killed.

Now, the increase of population and pollution and poverty are linked in this way (Fig. 16). In Japan we spent almost one trillion yen per year from 1970 on, in order to abate pollution and improve the environment. It needs money to

Fig.15. View of downtown Tokyo, west of Shinjuku area. Mt. Fuki can be seen behind the comulonimbus. August 28, 1989. (Courtesy of the "Asahi" newspaper.)

improve the environment: to abate air, water and soil pollution. If the country is poor, a large amount of money cannot be invested in antipollution equipment. If the pollution is very severe, then people cannot even get food from the fields since the soil is also polluted. Poverty and pollution are interrelated. Moreover, if the country is poor, one cannot invest large amounts of money, to treat the waste from each household, and a large population means large amounts of waste. If there is poverty, then the average life-span is reduced, and in order to protect your family, you have to raise more children. Therefore, population increases. In Japan, the population is now stabilized and we look forward to it slightly decreasing.

It is not enough merely to absorb pollutants. We have to consider how to reduce pollutants, and how to make use of

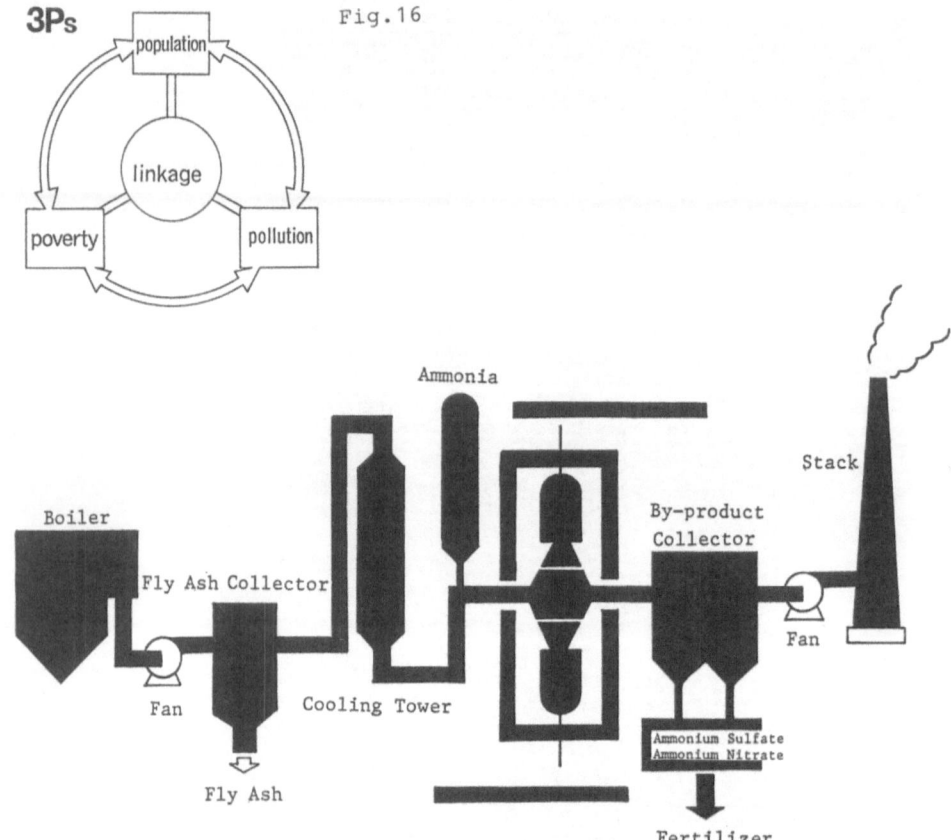

Fig.17. Flow sheet of electron-beam flue gas treatment

them, at the same time. Figure 17 is a flow diagram of flue gas treatment. By activating the molecules with an electron beam, and by a chemical process after this, we can produce fertilizers. This has the double effect of absorbing pollutants, and reusing pollutants as fertilizers. After burning oil, coal or other fuels, the smoke should be absorbed and passed through a tower. Then by chemically activating it by means of an electron beam, ammonium sulfate and ammonium nitrate are introduced to produce fertilizer. Actually, a test plant was built in the US.

We have to solve global environmental problems. As I said before, our scientists are supporting an international program as members of IGBP. We have to consider the whole environment systematically - not only the water; not only the atmosphere; not only oceanic pollution; not only soil pollution - we have to consider the whole Earth as a system. That is what the IGBP is going to study.

Two months ago we established in Kansai Science City a new institute called the Research Institute of Innovative

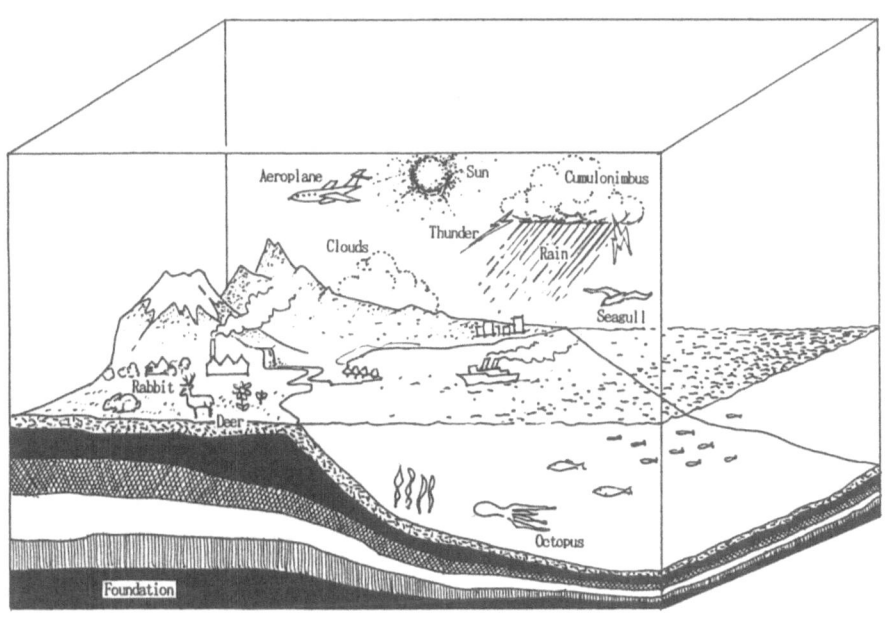

Fig.18. System analysis

Technology for the Earth or RITE. What we are going to explore are ways to improve the global environment. If you will open the brochure of RITE which I have distributed, you will see that we are going to find substitutes for fluorocarbon gases, and also we are going to make equipment which can solidify CO_2. With CO_2, it is rather easy to find the source. In the case of the automobile, the source is moving. But in the case of a generator of an electric plant, or the boiler of a factory the source is fixed. Therefore, we can easily find where the CO_2 is being emitted. There we can attach new equipment. We plan to look for devices to absorb CO_2, and even change it into methanol or some other useful material, in order to reuse CO_2 as a source of energy. Many other problems will be studied in the future. This is Japan's plan for the 21st Century. I am sorry I cannot go into detail here, but if you will look at the last part of my written paper, you will find more detail about RITE. In that paper, I discuss both sides: science and technology. In science, we have to work together with scientists from other countries to decrease uncertainty in the prediction of sea level and ecological changes. On the engineering side, as a high technology country, we have to find some technical means to decrease global changes. But it is also important to transfer technology to the developing countries. That means we have to devise new equipment or new machines which can be easily operated and do not cost very much.

Thank you very much for your attention.

Impact of Climatic Change on Agriculture from the Viewpoint of East Asia

M. Yoshino

Institute of Geoscience, University of Tsukuba, Tsukuba 305, Japan

Introduction

The planning of climatic impact studies after the first World Climate Conference in 1979 and the recent development of the World Climate Impact Studies Program (WCIP) have been described by Dooge (1990). As he pointed out, the effects of climate change on individual sectors of human activity and society have been the main subjects of study. The present paper reviews the results of studies related to climatic impact of global change. In particular, focus is given to studies relating to East Asia.

Development of the WCIP

The original plan of action consisted of the following four major program areas:
(1) Reduction of the vulnerability of food systems to climate;
(2) Anticipation of impacts of man-induced climatic change;
(3) Improvement in the science of climate impact studies; and
(4) Identification of climate-sensitive sectors of human activity.
    This plan of action for 1980-1983 was not carried out very successfully because of a lack of coordination between international and national organizations. In May, 1980, the Science Advisory Committee (SAC) was established to provide scientific guidance for the WCIP. The SAC revised the original plan of action and recommended a fifth program:
(5) Stimulation and coordination of climate impact studies.
    Following the first phase of the program, subsequent programs for each two year period from 1984-1985 to 1990-1991 have been prepared. At present the following four areas are going on:
(1) Greenhouse gases and climate change;
(2) Methodology and climate impact assessment;
(3) Coordination of activities in climate impact studies; and
(4) Monitoring of climate change and climate impacts.
    The main events in the development of the WCIP worldwide are shown in Table 1. Three main points are clear. They are: (i) national or regional levels have been stressed in the project, (ii) climate impact assessments have been summarized, and (iii) effects of global warming on human activity and society have been the main concern.
    The development of the WCIP in Japan started gradually in the early 1980's. Its chronology is roughly summarized in Table 2.

The Global Environment
Editors: K. Takeuchi · M. Yoshino © Springer-Verlag Berlin, Heidelberg 1991

TABLE 1

Development of the WCIP in the world:

Year	Subjects
1979	First World Climate Conference. Working Group on the "Impacts of Climatic Change of Variability on Society".
1980	Feb: Ad hoc group of experts for WCIP. May: Governing Council of UNEP. Initial Plan of Action for 1980-1983.
1981	Science Advisory Committee, SAC, (WCIP and WMO) first meeting.
1983	A compendium of climate impact projects at national level was distributed.
1985	Assessment Conference on Greenhouses at Villach, Austria.
1985	Kates, R. W. et al.: Climate Impact Assessment. SCOPE 27, 625p. This monograph was a result of "Project on Climate Impact Assessment" of the Scientific Committee on Problems of the Environment of ICSU (SCOPE).
1985	SAC concerned the National Climate Impact Activities.
1986	WCIP meeting in Warsaw, Poland. Nine countries reported their activity.
1988	Parry, M. L.: The Impact of Climatic Variations on Agriculture. Kluwer, Dordrecht, Vol. 1, 876p. and Vol. 2, 764p.
1989	March: International Networkshop on Climate-Related Impacts in Boulder, USA. Five countries presented report.

The Japanese Study Group for the WCIP and WCAP (World Climate Application Program) had their first meeting in Tokyo on 26th October, 1983. The Group publishes an inventory of researchers whose working fields are related to the WCIP and WCAP. This inventory contained details of 202 peoples in 1984. In 1990, about 260 active members are included. Meetings are held in Tokyo twice a year. Among many study fields, the impact on crops, livestock and agricultural production are the strongest. Next come studies on past climate, climatic fluctuation and climatic changes. The third group, which have relatively high potential, are the impact of climatic variability or change on natural ecosystems and wildlife, on human society and industries, and on cryosphere and oceans including fisheries.

TABLE 2

Development of the WCIP in Japan:

Year	Subjects
1980	Y. Fukushima stressed the importance of WCIP in Journal of Agricul. Met. in Japan.
1981	"Climate-Related Problems Panels" in Japan Meteorological Agency (JMA).
1981	"Office of Climate Program" in JMA to establish counter-measures for climatic change.
1982	Investigation on the relationships between weather/climate and local industries for six regions in Japan by JMA.
1983	First meeting of the Japanese Study Group for WCIP and WCAP.
1984	Inventory of WCIP and WCAP in Japan.
1985	National programs of WCIP and WCAP in Japan.
1985	"Study group on meteorological services for industry" organised by IMA and about 100 companies.
1987-1988	Studies in association with the Project on "Impact of Climatic Variations on Agriculture by the IIASA.
1987-1988	Project study on climatic change in the coming half century and estimation of food, energy and water in Japan: Leader M. Yoshino.
1988	Monographs entitled "Problems in climatic impacts and application studies in Japan.
1989-1990	Project study on the climate in the coming half century and agriculture, forestry, fishery and human environment in Japan: Leader M. Yoshino.

In July 1988 the Japanese Study Group for WCIP and WCAP published a monograph entitled "Problems in Climatic Impacts and Application Studies in Japan" (231 pages) in Japanese (Yoshino, 1988 a). Other monographs have been the result of project studies on climatic change in the coming half century and an estimation of food, energy and water in Japan (Yoshino, 1988 b and 1990 c). In the latter studies, a scenario postulating a temperature increase of 2°C, (2.5°C in North Japan, but 1.5°C in South Japan,) and a doubling of CO_2 was considered. Temperatures of 2.5°C in winter and 1.5°C in summer were proposed. For precipitation, three cases were proposed: more, normal and scarce. The final results of these project studies are published in 1991.

Regional Aspects of Warming

Records of global and hemispheric temperatures for the period
1861-1988, show that they were roughly constant in cycles of
11-12 years until about 1920 and since then, they have
gradually increased with a peak around the beginning of 1960,
(Jones, 1989). In 1980 there was a particularly striking
increase in both hemispheres. In general, the increasing rate
can be recognized roughly as 0.5°C per 100 years since the
beginning of the 20th century. This figure coincides with the
value estimated by model simulation for CO_2.

The regional aspects of warming in East Asia are shown
first, using model simulation. In Figures 1 - 4, results of
simulations by various General Circulation Models are shown
for surface air temperature (°C) and precipitation (mm/day) in
the case of (2 x CO_2) in winter (mean of December + January +
February), and (1 x CO_2) in summer (mean of June + July +
August). The models used are GISS, GFDL, MRI (SST), UKMO, OSU
and MRI. In winter, surface temperatures show a clear
increase of several degrees in the northern part of East Asia,
but 2°-3° in most parts of South and Southeast Asia. In
summer, the contrast between north and south in monsoon Asia
is not great, except with the UKMO and MRI models. The
results of precipitation simulation show relatively large
differences between the models, particularly in summer. In
tropical areas, the differences are striking. It is certainly
necessary to discover further details for regional
consideration, especially for agriculture and forestry in the
case of doubling CO_2. Zhao (1989) studied climatic change in
China induced by doubling CO_2, as simulated by the five GCMs.
His results for the regions in China coincide with those in
the present report.

Yamamoto (1990) calculated the rate of air temperature
change for various regions. Based on his results, the changes
of air temperature over the Northern Pacific Ocean, averaged
for every 20° latitudinal zone, are reproduced in Figure 5.
The differences between latitudes are very striking. There
are several important findings:
(1) The general tendency during the last 80 years differs in
different latitudes.
(2) The range between maximum and minimum is largest in the
zone 70°-50°N and smallest in the zone 30°-10°N.
(3) From the beginning of this century, a warming tendency
is evident in the zones 70°-50°N and 50°-30°N, but not in
lower latitudinal zones.
(4) Since 1970, data show a warming tendency in the zones
70°-50°N and 10°N-10°S but cooling in 50°-30°N and no clear
tendency in 30°-10°N.
(5) Between 1940 and 1970, roughly speaking, there is no
striking tendency between the zones.
In conclusion, it is clear that the regional differences
are striking, and that the air temperatures over the Pacific,
50°-30°N, a region closely connected to East Asia, have
decreased during the last 10-15 years.

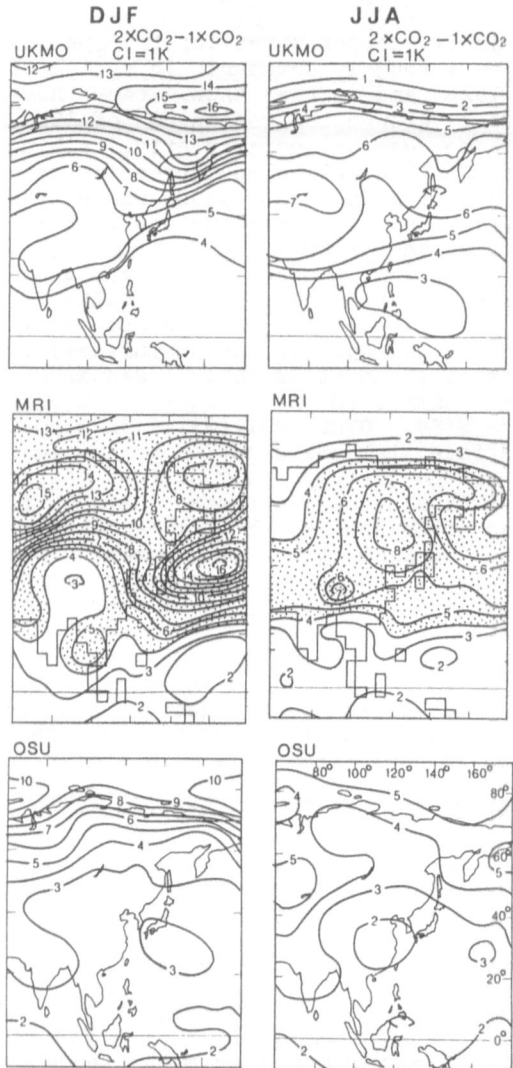

Figure 1:   Results of simulation of surface temperature
(degrees K) by various GCMs for winter (DJF) and summer (JJA)
in the case of doubling CO_2 (JMA, 1990). MRI is the model by
Meteorological Research Institute in Japan, MRI(SST) is by
MRI, but introducing the estimated zonal mean SST rise by
Manabe et al., (1988) for the purpose of regional test, UKMO
is by the UK Meteorological Office in England, GFDL is by the
Geophysical Fluid Dynamics Laboratory, NOAA, USA, GISS is by
the Goddard Institute of Space Science, USA, and OSU is by the
Oregon State University, USA.

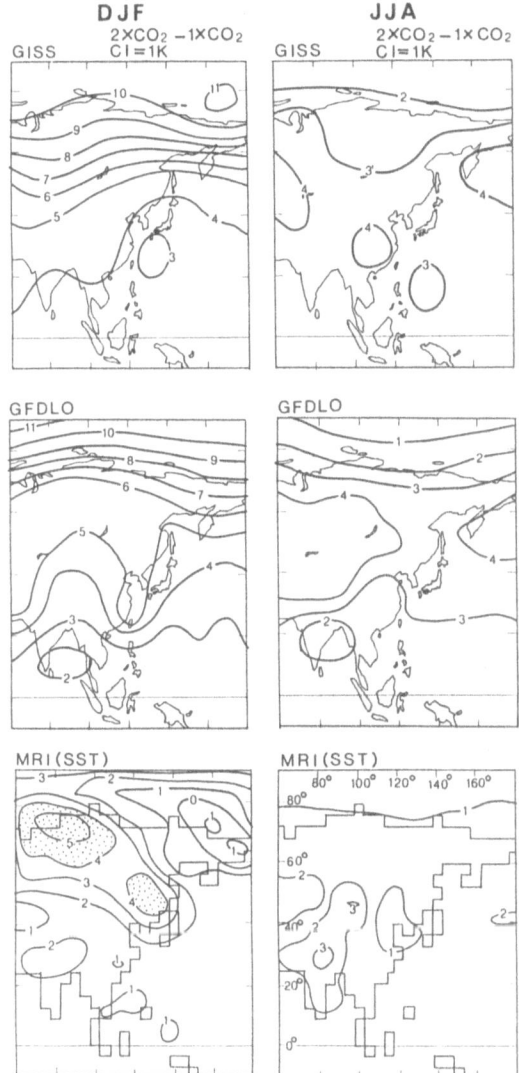

Figure 2:   Continuation of Figure 1.

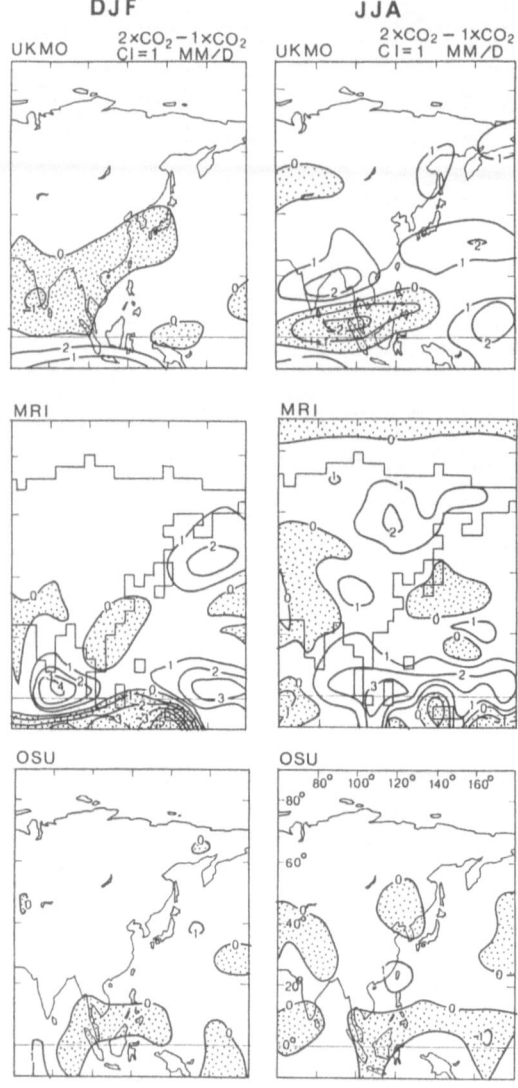

Figure 3:   Same as Figure 1, but for precipitation (mm/day).

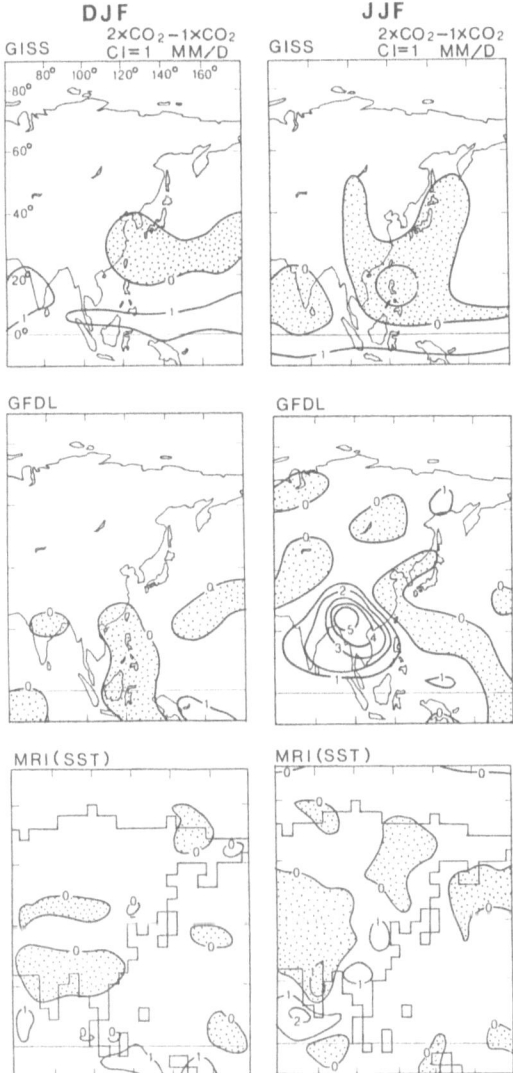

Figure 4:   Continuation of Figure 3.

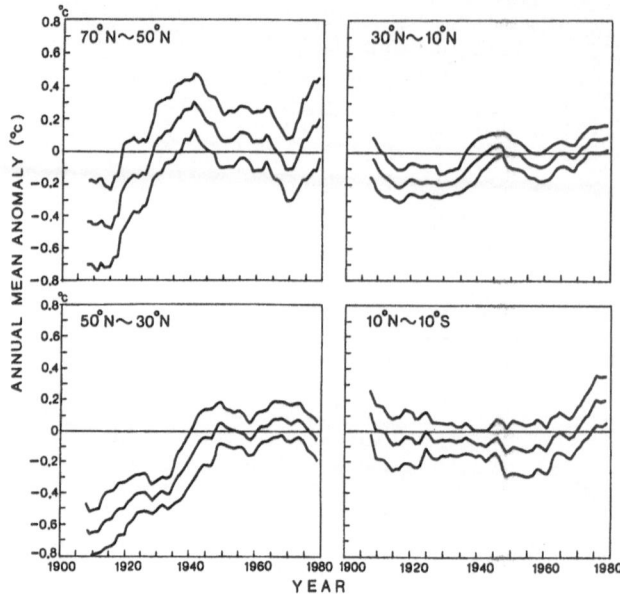

Figure 5: Secular change of air temperature over the Pacific Ocean averaged for the 20° latitudinal zones (adapted from Yamamoto, 1990). Curves show 15-year-running mean and range of its 95% reliability.

Impacts of Warming on Agriculture

Effective accumulated temperature (accumulated daily mean air temperature higher than $10°C$) will shift present levels in the southern part of Tohoku District, Northeast Japan, to the northern part of Tohoku District or to the southern part of Hokkaido. The condition in the southern part of Kyushu moves to the Kanto Plain. Essentially, temperature climates will move about 4° northwards. As a result, it should be possible to cultivate subtropical or marginal tropical crops in the southern part of Kyushu. It may be necessary to introduce Indica-type rice or hybrid of Indica- and Japonica-type in these regions.

Table 3 shows changes in rice yields in North Japan in the case of doubling CO_2, as estimated by Uchijima, (1988). In Hokkaido and Tohoku, Northern Japan, air temperatures in summer will increase about 3° based on estimations by the GISS model, and the upper limit of paddy fields will become higher by about 500 m. But, it is important to recognise that paddy production will decrease, if the present variety continues to be cultivated. If a middle-ripening variety or late-ripening variety is introduced, production will increase. Generally speaking, it is thought that a 1°C increase in mean annual air temperature would tend to advance the limit of cereal cropping in mid-latitudes of the northern hemisphere by about 150-200 km, and raise the upper limit on mountain slopes of arable agriculture by about 150-200 m (Parry et al., 1990). The figures in northern parts of Japan roughly coincide with these values.

TABLE 3

Change of air temperature and rice yield in North Japan in the case of doubling CO_2 :

District	Hokkaido	Tohoku
Average monthly air temperature July + August (°C)	+3.5	+3.2
Accumulated Air Temperature (%)	+35	+29
Upper Limit of Paddy Cultivation (m)	+556	+508
Yield Index of Paddy (%)	+5	+2
ESTIMATED PADDY PRODUCTION:		
Present Variety (%)	-7	**
Middle Yield Variety (%)	+23	**
Late Yield Variety (%)	+26	**

** Average of paddy supply for Tohoku district is +7%
+ Sign denotes increase or higher up, and
- sign decrease or lower down.

Climatic yield capability in China was calculated (Chen et al., 1990), showing an increase of about 10% due to doubling CO_2, with a maximum increase of 20% in the inner part of South China. It is noted that the growing regions in China may be moved northward about 5° of latitude generally. However the Chinese hope to make further detailed studies.

In the report of Working Group 2, IPCC, Chapter 2 reviews thoroughly the potential impact of climate change on agriculture and forestry (Parry et al., 1990). It points out that present-day vulnerability to climate is one of the most important problems. On average, 63% of the land area of developing countries is climatically suited to rain-fed agriculture. But, it is believed that this figure reaches about 84% in Southeast Asia (FAO, 1984).

Direct effects of elevated CO_2 and other greenhouse gases have been experimentally studied. Here the main questions are photosynthesis, respiration and growth. These are different according to two groups of plants: C3 (eg, wheat, rice, barley, root crops, legumes) and C4 (eg, maize, sorghum, millet, sugar-cane). It has been estimated experimentally that the average effect of doubling CO_2 on wheat grain yield is an increase of 35%, but that maize shows no significant increase. The other problems to be considered are: plant development, yield quality, stomatal aperture, water use and biological nitrogen fixation. These are not mentioned in detail here because our knowledge of these problems is too generalized to discuss quantitatively conditions in Monsoon Asia.

According to Zhang´s (1989) estimates, yields of rice, maize and wheat will increase by about 10% overall nationally, but there may be modest decreases in the north and east under conditions of 1°C warming, with precipitation increases of 100 mm. If there is no increase in available moisture, however, maize yields in the east and central regions may decrease. These temperature and moisture conditions depend basically on monsoon activities in summer and winter. So the problems in China and Japan, as well as in the other parts of Monsoon Asia, are closely related to the future behavior of monsoon activities.

In conclusion, the present questions are: (i) regional climate scenarios, which are closely related to monsoon activities, (ii) the capability of farmers to respond to climate change, and (iii) the tolerance of crops to changing climatic conditions. Further, changes in land use are important; (i) changes in farmed area, (ii) crop type, (iii) crop location and (iv) crop calendar under new climatic conditions should be studied. Changes in management such as irrigation, fertilizer use, control of pests and diseases, soil drainage and erosion, among others, will be considered in relation to the questions and problems mentioned above.

Sea Level Rise

Global mean sea level rise was about 10 cm during the last 100 years, according to Gornitz et al., (1982). This rise was caused mainly by heat expansion of sea water and melting of land ice such as mountain glaciers. If we estimate 1.2 - 3.0°C of global warming, mean sea level rise in about 2030 will be 5-17 cm by heat expansion of sea water, 4-9 cm by melting of the Greenland ice sheet, and 3-24 cm by contribution of the Antarctic ice sheet (Japan Met. Agency, 1990). A review of sea level rise on the coast of China based on these results, is given in this part of the study. Also, a comparison with some evidence from other parts of East Asia is attempted.

Huanghe River Delta

The Holocene Huanghe River (Yellow River) Delta is one of the largest deltas in the world. It has an area of 25,000 km², (which is next largest after the Mississippi delta with an area of 50,000 km²). It consists of three younger deltas; the delta near Tianjin, the present-day delta and the abandoned delta.

In the Tianjin delta, the area lower than 4 m above sea level is considered to be vulnerable to sea level rise. From south to north, this area is more than 150 km and from to east to west, it is about 50-60 km; in total the area is about 8,000 km².

The modern delta with an area of 5,400 km² has rich oil resources. Crude oil production has been increasing rapidly

TABLE 4

Relative sea level variation in the recent 20-30 years in the Huanghe River (Yellow River) and its adjacent areas:

Name of Tide Gauge Station	Latitude	Longitude	Observation Period	Mean Velocity (cm/year)
Huludao	40° 42´N	120° 54´E	1955-1981	+0.19
Qinhuangdao	39° 54´N	119° 30´E	1956-1980	+0.21
Yingkou	40° 36´N	112° 12´E	1952-1971	+0.11
Tanggu	39° 00´N	117° 36´E	1950-1981	+0.81
Yangjiaogou	41° 06´N	119° 54´E	1952-1978	+0.19
Longkou	37° 36´N	120° 12´E	1961-1981	+0.25
Yantai	37° 30´N	121° 24´E	1953-1981	0.00
Rushankou	36° 48´N	121° 30´E	1960-1981	0.00
Qingdao	36° 00´N	120° 18´E	1952-1980	0.00
Shijiusuo	35° 24´N	119° 30´E	1968-1981	0.00
Lianyungang	34° 36´N	119° 06´E	1953-1981	-0.54
Wusong	31° 30´N	121° 24´E	1950-1971	-0.95
Luhauashan	?	?	1963-1981	0.00

+ Rise
- Subsidence

(Data source: Wang et al., 1986)

in the delta, reaching 32,000,000 ton in 1987 or 23.6% of the country's total production. This delta is considered to have an observable future economically.

The abandoned delta with an area of 7,100 km^2 was built up during the period 1128-1854, when the Huanghe River flowed into the Huanghai (Yellow Sea), changing the river course from northern Jiangsu. Population density is high in the area and this an important agricultural region.

Records from the tidal stations located along the Bohai show a marked relative rise in sea levels. As shown in Table 4, the maximum sea level rise has been at a rate of +0.81 cm/year at Tanggu, for the last 30 years (1951-1980). If we calculate the rate at the North Paotai Station near Tanggu for a longer period (1910-1979), it is +0.43 cm/year. This area is a center of land subsidence due to the heavy pumping up of ground water. According to a careful survey of land subsidence in this area, the amount of subsidence was 0.70 cm during the period 1968-1974 and the relative sea level rise was +0.735 cm in the same period. In addition, it should be noticed that this relative sea level rise is remarkable only at the stations located on the coast of Bohai. Figures from the stations on the coast of Shandong Peninsula are stable, showing 0.00 in Table 4. At the stations on the coast of middle part of China, for instance, Wusong on the mouth of Changjiang (Yangtze River), the relative sea level rise is negative, indicating a descending sea level. The reason why the tendency should be opposite, is not clear. It should be stressed that the rate of relative sea level rise differs locally due to geomorphological conditions and human activities such as ground water draw-off.

The following figures were observed during the period 1956-1980 (Okada, 1988). The crustal movement is 0.0 - 0.5 cm/year (subsidence) on the coast along the Pacific (in some parts of Tohoku District and Hokkaido, North Japan, there is 0.0 - 1.0 cm/year subsidence), while there is a 0.0 - 0.25 cm/year uplift on the coast along the Japan Sea. These values should be considered in relation to global sea level change. In addition, the change of velocity and positional shift of the Kuroshio current, a subtropical current in the North Pacific, has an effect on tide levels on the coast, because the Kuroshio current is about 1 m lower on its coastal edge than on the opposite side. So, it is considered that this will have an effect on the sea level change in Southwest Japan and probably also in Taiwan.

The major direct impacts of relative sea level rise on the Huanghe River Delta are submergence and an increase in salinity, (Ren, 1990). The winter monsoon and typhoons requently cause strong winds and high tides. In the hundred years before 1949, serious situations occurred seven times. Sea water invaded the land 15-50 km. At Tanggu, sea level increased by more than 4.6 m above the normal level on August 31, 1939, 5.28 m on August 16-20, 1985, and 6.1 - 7.0 m on April 28-29, 1895. The height of coastal banks is normally about 4 m. Therefore, these values mean catastrophe in this area with an increasing tendency towards damage both in frequency and amount.

The relative sea level rise and the lowering of ground water levels result in the greater effect of sea water on ground water in the coastal plain. Pollution by sea water can be unavoidable. Economically this area is a center of sea salt production, 2,000,000 ton/year. Because this is a major rice production area, shortage of good irrigation water, difficulty of drainage, and effect of salty ground water are serious problems for cultivation.

Summarizing the phenomena mentioned above, impacts of relative sea level rise are as follows: (i) severe damage was caused mainly by winter monsoons and typhoons, and occasionally by cyclones in April; (ii) sea level during the worst cases increased 6-7 m above normal and was regularly 4-5 m higher than normal; (iii) sea water invaded 40-50 km inland in the worst cases; (iv) the effect of sea water intrusion into the inland ground water will be increased; (v) drainage will become more difficult; (vi) rice cultivation, sea salt production, oil production, and port construction will suffer severe damage.

Chanjiang (Yangtze River) Delta and Northern Coast of Jiangsu Province

Sea level rises on the Changjiang (Yangtze River) Delta and on the northern coast of Jiangsu Province and their relation to climate change and impacts have been looked at in a number of studies, (Long et al., 1989; Xu et al., 1989; Zhu et al., 1989). According to their results, the rate of land subsidence is 0.16 cm/year on the coastal plain of Northern Jiangsu Province, 0.41 on the Changjiang delta, 0.50 in Shanghai and 0.11 on the coastal plain around Hangzhou.

Sea level rises and their effects on tidal levels, coastal construction and morphology, human society and economy are shown in Table 5. This table indicates that the most serious impacts will occur in the region of Changjian River mouth and delta, and on the coastal plain around Hangzhou. The distribution of degree of damage suffered because of sea level rise on the coast of Jiangsu Province is shown in Figure 6. In this figure, the potential productivity of medium-variety rice as calculated by Long et al. (1989) is also shown. Wheat and maize potential production have a similar regional tendency. Based on this figure it is concluded that the several areas of severe impact, such as coastal erosion, river drainage, wind and tidal damage to coastal construction, flooding over marshy lands, and saline water intrusion, should be considered. The quantitative impacts on agricultural production, region to region, can be estimated for various crops.

TABLE 5
Sea level rise and its effects on the tide level, coastal morphology, human society, economy etc.:
(compiled from a table by Zhu et al., 1989)

	Huanghe River Delta
Estimated sea level rise in 2025	18-28 cm
Maximum difference between high and low tide	4 m
Observed record of high tide	3.28 - 3.51 m above normal
Population density	355 - 524 head/km^2
Economic density	70,000 - 90,000 US$/km^2
Impact of sea level rise	1. Coastal erosion, 2. Destruction
Counter-measure	1. Coastal protection forest

* Tamaki, M. (1988): Kankyo-gijutsu, 17, 696-704.
** Onozaki, T. and Ogura, N. (1987): Japan Limnological Society, 52nd Annual Meeting Abstracts, 43.

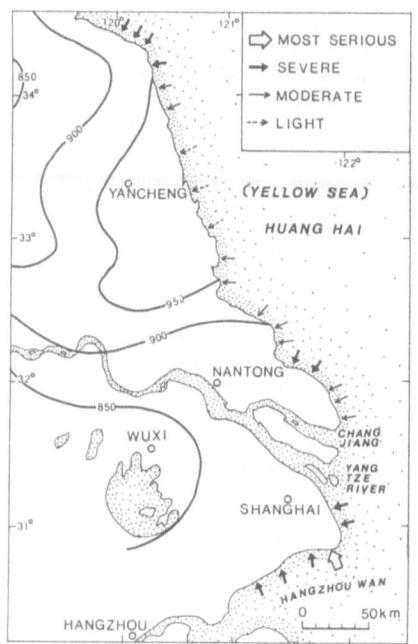

Figure 6: Grade of damage suffered from the effect of sea level rise on the coast of Jiangsu Province and the potential productivity of middle-ripening variety of rice (kg/mu = 15 kg/ha).

Coastal geomorphic response to future sea level rise and its implication for the low-lying areas of the Bangkok metropolis was studied by Somboon (1990). He has pointed out that a combination of subsidence and sea-level rise will seriously affect residential areas, pollution of surface water and groundwater, flooding, waste water drainage and treatment, agricultural land, and industrial and commercial activities in the Bangkok metropolis area in the future.

It is concluded, therefore, that similar problems can be expected in many delta regions in Southeast and East Asia.

Acid Rain

In Japan:

The Environmental Agency of Japan, has made observations of acid rain since 1984, and has studied the effects of acid rain on inland waters at 133 lakes, and on soil as well as on plant growing at 12 stations. The Japanese Government reported on August 15, 1989, that a harmful rain affected by nitrogen dioxide (NO_2), sulfur dioxide (SO_2), and other pollutants from factories, electric power plants and cars, has been falling on Japan. In Figure 7, the spread of acid rain (pH) in Japan is shown. Generally speaking, lower (more acid) pH values were observed in the southwestern regions of the Japan Sea side of

```
 1 : 4.6 Iwamizawa
 2 : 5.0 Sapporo
 3 : 5.0 Yufutsu Mukawa
 4 : 4.9 Aomori
 5 : 4.8 Hirosaki
 6 : 4.7 Sendai
 7 : 4.9 Okawara
 8 : 4.4 Musashino
 9 : 4.8 Tokyo Koutouku
10 : 5.0 Toyama
11 : 4.8 Kosugi
12 : 4.5 Ishikawa Tatsukuchi
13 : 4.6 Yoshinoya Ishikawa
14 : 4.6 Nagoya Meitouku
15 : 4.5 Osaka
16 : 4.6 Ikeda
17 : 4.6 Kyoto Keihoku
18 : 4.5 Kyoto Yasaka
19 : 4.9 Shobara
20 : 4.5 Hiroshima
21 : 4.5 Yamaguchi
22 : 4.9 Hagi
23 : 4.5 Kochi Kahoku
24 : 4.6 Ochi Kochi
25 : 4.5 Omura
26 : 4.7 Nagasaki
27 : 4.6 Yaku Kagoshima
28 : 4.6 Kagoshima Tachutake
```

HOKKAIDO

HONSHU

SHIKOKU

KYUSHU

0   200 km

Figure 7:   Spread of acid rain   (pH) in Japan (Japan
Environmental Agency, 1989).

Kyushu and Honshu.   The reason why the southwestern region of
Honshu   shows stronger acidity,   is probably   the recent
industrialization of China and South Korea.   The effect of
westerly winds in transporting pollutants from these up-stream
source regions should be studied, because stronger acidity was
observed in winter, when the westerly in the troposphere is
stronger in East Asia.   Secondly, in Figure 6, lower pH values
are shown in the heavily industrialized regions such as around
Osaka and Tokyo.   The strongest acidity, a pH of 4.4, was
measured in Tokyo, which is slightly better than the worst
place-average, 4.0, in Europe and North America.

In Table 6, changes in annual mean pH of rain water
measured at two stations in Japan are presented.   At Ayasato
in northeast Japan, the pH level was slightly high, but it has
shown practically no change since 1978.

The observed values of pH in Yokkaichi are not shown in
Figure 7, but, according to measurements from 1961 to 1979,
the average has dropped below 5.6 (acid rain) since 1963, and
4.4 (heavy acid rain) since 1968. This has been attributed to
the development of a petrochemical complex in Yokkaichi City.
According to Taniyama (1990), the effect of air pollution
including acid rain on the growth and yield of rice in the
Yokkaichi Region were as follows: (i) effect of SO_2 was not

TABLE 6

Secular change of annual mean pH of rain water in Japan:

Year	Ayasato, Iwate Prefecture, North East Japan*	Fuchu, Tokyo, Central Japan**
1976	5.2	-
1977	5.0	4.18
1978	4.7	4.30
1979	4.9	4.36
1980	4.9	4.29
1981	4.8	4.36
1982	4.7	4.41
1983	4.6	4.39
1984	4.5	4.28
1985	4.8	4.19
1986	-	4.28

* Tamaki, M. (1988): Kankyo-gijutsu, 17, 696-704.
** Onozaki, T. and Ogura, N. (1987): Japan Limnological
Society 52nd Annual Meeting Abstract

significant, but SO_3 (mg/day/1 m) has a close relationship to
the rice yield; (ii) since 1972, the air pollution improved
and its effect on rice production changed for the better.
Changes of rice yield in Yokkaichi City, where the air
pollution is more serious and in the whole Mie Prefecture
(whole are is 5,768 km², and paddy fields are about 62,000 ha)
are given in Figure 8. It is clearly seen in this figure that
the yield in Yokkaichi became strikingly lower than the
average in Mie Prefecture after 1958, when the operation of
the petrochemical complex began in Yokkaichi. It is also
clear that the centralized smoke stacks of 150-200 m caused no
further serious damage, with a one year lag and, after the
regulation of total emissions of sulfur oxides, the yield
stabilized at a higher level, mostly reaching the average of
Mie Prefecture. Taniyama obtained an experimental equation
between the amount of sulfur oxides, x (t/year), and yield
decrease of brown rice, y (kg/ha), in Yokkaichi City from 1971
to 1981;

$$y = 3.3x + 121$$

The total cost of decreased rice yield from the beginning of

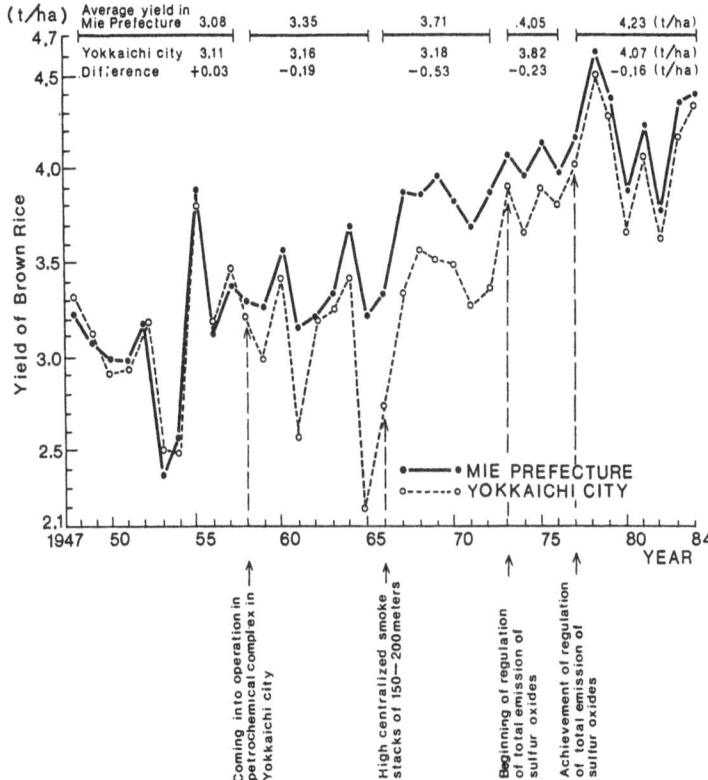

Figure 8: Secular change of yield of brown rice in Yokkaichi City and Mie Prefecture, 1948-1984, and some chronological evidence (mainly by Taniyama, 1990; slightly changed by Yoshino).

the operation of the petrochemical complex in the Yokkaichi City area in 1958, until 1984 is estimated as more than 65 million US dollars.

In China and Korea

In September and October, 1985, an observation on the acidity of rain water was made in Chongqing, Sichuan Province, in China. As is shown in Table 7(a), the mean value of pH is 4.06 in the city center and 4.31 in the suburbs. It is extraordinarily high acidity. In particular, when rainfall intensity is less than 0.5 mm/hour, the mean pH is 3.66. When it is in the range of 0.5 - 2.0 mm/hour, the mean pH is 3.98. So in situations of weak rainfall intensity, there is strong acidity. In the suburbs (airport), the mean pH is slightly higher, but it is still less than 4.00, indicating strong acidity.

In the Shanghai region, an observation of acid rain was made in June and July, 1986. The mean pH values and their

TABLE 7(a)

Examples of measured pH of rain water at several points in China:

Stations	Chongqing[1] Inst. Envir.	Airport	Shanghai[2] City Center	Airport
Latitude	29° 33´N	29° 28´N		
Longitude	106° 31´E	106° 21´E		
Observation period	September -October 1985	September -October 1985	June-July 1986	June-July 1986
Mean pH	4.06*	4.31*	5.08* 4.70**	6.19* 4.86**
Range of pH	3.08 - 5.18	3.27 - 5.77	3.24 - 7.69	3.54 - 9.80

* Mean of all cases,
** Mean of cases pH<5.6.

Data source:
[1] Huang, M.-y. et al., 1988.
[2] Shen, Z.-1. et al., 1989.

TABLE 7(b)

Rain water pH in the city of Pusan, South Korea:

Observation Period	March- September 1983	January -June 1986	March -June 1988
Number of observation points	8	11	10
Number of observed rainy days	26	19	13
Range of mean pH	4.4 - 6.6	4.4 - 5.4	4.6 - 5.8
Mean pH of all cases	5.4	5.1	4.9
Lowest observed record pH	3.8	4.1	3.3

Data source: by courtesy of Prof. S. E. Moon, Department of Atmosphere and Science, Pusan National University.

range are given in Table 7(a). In the Shanghai region, the ranges are quite wide, but mean pH is above those in Chongqing. In the suburbs of Shanghai, the relationship between rainfall intensity and acidity is not linear. Namely pH is a low 5.40, in the case of 1 - 2.5 mm/hour and, in the cases of weaker intensity and stronger intensity rain, pH values are greater (more than 6.00). It is considered that the effect of pollution on acidity relates to the size and number of rain drops, and is weaker in times of weaker and stronger rain than it is in the case of an intensity of 1 - 2.5 mm/hour.

In Pusan, South Korea, conditions are similar to those in Shanghai; the range of mean pH observed at 8-11 points shows 4.4 - 6.6, with low values of 3.3, as shown in Table 7(b).

Global Change and Agriculture, Forestry and Fishery

Fluctuation According to the Historical Development of Agriculture:
An important problem in agricultural production in the future, along with global change, is not only total production, but also year to year fluctuation, due to varying natural and socio-economic conditions. The fluctuation of paddy rice yield in Japan is indicated by the yield results for the last 100 years.

The long term trend of paddy rice yield in Japan, can be divided into three periods: Period I started from the end of last century and lasted until 1910-1915, when the trend showed a gradual increase. Period II, 1915-1945, has no striking tendency, but remained a roughly constant yield. Period III, since 1945, shows marked increase. In Figure 9, the example of North Japan, including Hokkaido and Tohoku District (except Fukushima), is given. It can be seen that larger fluctuations occurred in the second half than in the first half of each period. The standard deviation in each period, as given in Table 8, becomes one order smaller in Period III than in Periods I and II. This may be attributable to technical developments in rice cultivation, introduction of new varieties, and so on. It is, however, quite interesting to note that the second half of each Period shows strikingly greater fluctuations than the first half, as mentioned above. This is of course not related to climatological fluctuations, but may be related to the character of technological developments in agriculture. Detailed discussion is omitted here because of space, and is given elsewhere (Yoshino, 1990

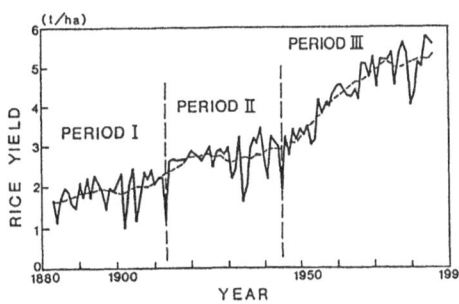

Figure 9: Secular change of rice yield in North Japan for the period 1883-1986.

TABLE 8

Fluctuation (defined by standard deviation) of Paddy Rice
Yield in Each Period in North Japan:

Period	Years	Standard Deviation* (t/ha)
Period I - first half - second half	1885 - 1917 1885 - 1900 1901 - 1917	 0.118 0.244
Period II - first half - second half	1918 - 1950 1918 - 1933 1934 - 1950	 0.122 0.168
Period III - first half - second half	1951 - 1984 1951 - 1967 1968 - 1984	 0.063 0.081

* Root-mean-square of deviation of each year from the 11-year-running mean.

b). If these characteristics appear in other crops, we should look seriously at future agricultural production.
Uchijima (1981) expressed variability in crop yield thus:

$$Y(t) = YT(t) + Y'(t),$$

where $Y(t)$ is the observed actual crop yield (t/ha), $YT(t)$ is the non-linear trend of crop yield (t/ha), and $Y'(t)$ is deviation of crop yield from the non-linear trend (t/ha). In his study the entire period from the 1880s to the 1970s was not divided into sub-periods. He also calculated the yield index, $IY(t)$, as

$$IY(t) = \frac{Y(t)}{YT(t)}$$

and the coefficient of variance, CV_{IY}, of the time series of yield index as

$$CV_{IY} = \frac{{}^{\circ}IY}{IY} \times 100 \ (\%)$$

where IY is the serial mean of time series of yield index. CV_{IY} was 9% for rice, 12% for barley, 14% for wheat, maize or millet and 16% for rye. The reason that the coefficient of variance of paddy rice is the smallest in Japan can be attributed to the development of irrigation systems, which soften the adverse effects of unfavorable weather conditions.
In conclusion, the fluctuation of yields is affected by climatic fluctuation, but sits magnitude differs in accordance with the development of agricultural technology. In a long

period, it becomes generally smaller because of the development of technology. But, if we look at it subdivided into terms such as 15-20 years, fluctuation is greater in the later terms. This may be caused by the step wise development of human response including agricultural systems.

Flow of Impacts

Summarizing the impacts of recent global environmental change caused by the human activities on agriculture, forestry and fishery, a tentative flow chart is shown in Figure 10. The striking acceleration of human activities during recent years, including population increase, is the starting point of the flow. It results in "Environmental Change I". Warming by the greenhouse effects, a major part of "Environmental Change I" causes "Environmental Change II". Some parts of "Environmental Change I" directly cause "Environmental Change II". In "Environmental Change II", photo synthesis is considered as correlated with the changes in atmospheric, oceanic and soil environments. Photosynthesis is a biological process in plants, so it should be regarded differently for agriculture, forestry and fishery. Further consideration is needed on this point.

Impacts of "Environmental Change II" finally result in the last box containing many factors in agriculture, forestry and fishery. These results feed back to the initiating processes, which are expansion of arable lands, destruction of natural vegetation, change of land use, speed-up of desertification on the land, oceanic pollution, destruction of the marine ecosystem and destruction of coastal environments.

The problems shown in the last box are particularly important in East Asia. But, unfortunately, the quantitative analysis is yet to be done.

Summary and Conclusion

International and Japanese activities related to WCIP have been reviewed first in this paper. Secondly, the regional aspects of global warming were dealt with and potential consequences of observed records were computed by the GCMs. It was stressed that regional differences are striking; in particular, it is of importance for East Asia that air temperatures over the Pacific have been decreasing over the last 10-15 years.

It appears that the following studies should be made in relation to the impact of warming on agriculture: (i) the development of regional scenarios which are closely related to monsoon activities in the area studied; (ii) the capability of farmers to respond to climatic change; the tolerance of farmers, and the tolerance of crops; (iii) further, farming areas, crop types, crop locations and crop calendars should be considered.

The questions related to sea level rise are these: (i) severe damage occurred mainly during the winter monsoon and typhoons, but occasionally during cyclones in April; (ii) sea level during the previous worst cases rose 6-7 m above normal and sea water invaded 40-50 km inland-wards; (iii) sea water

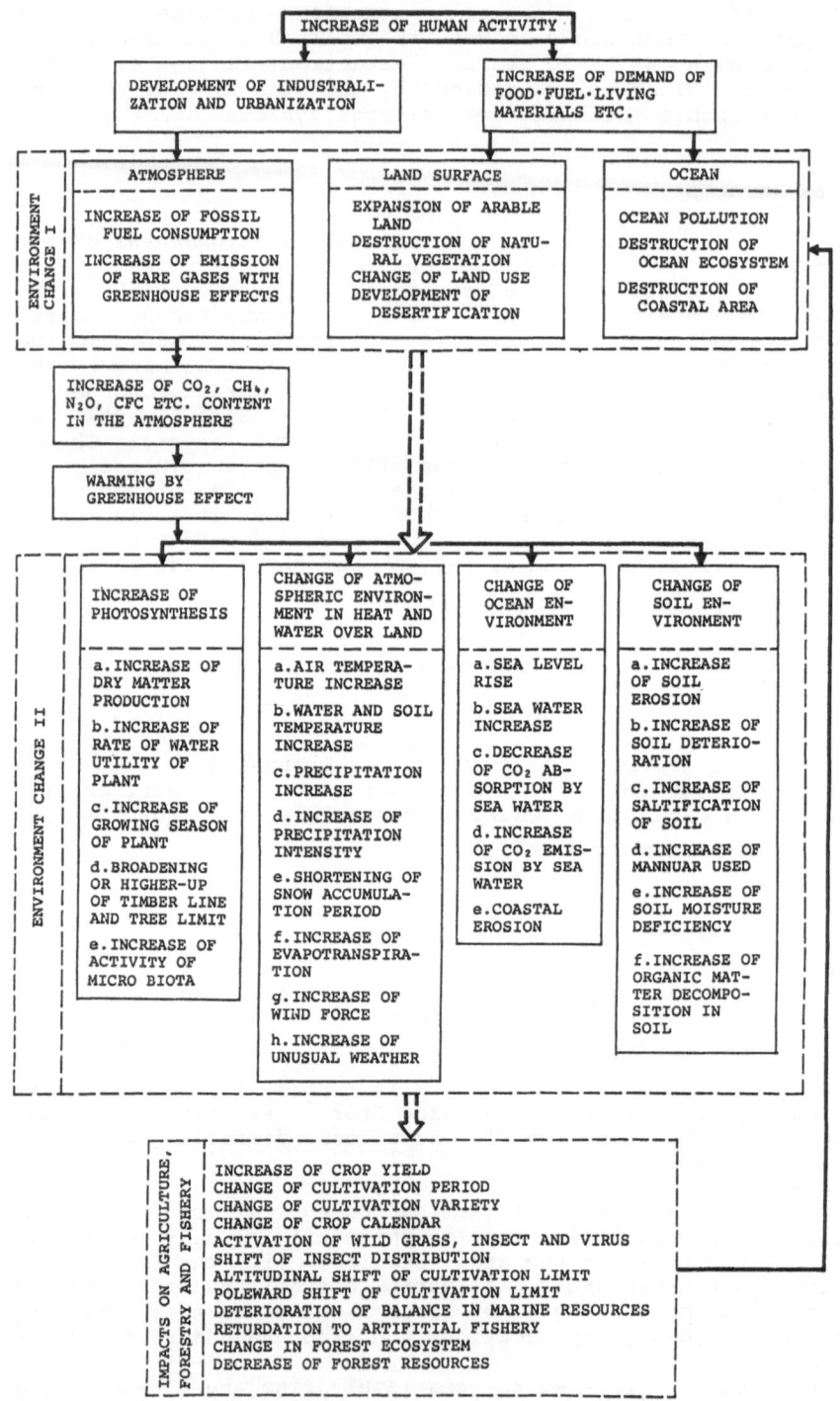

Figure 10: Flow chart of impacts or interrelationship of human activities to agriculture, fishery and forestry through the environmental changes.

intrusion into inland ground water and difficulties of drainage are likely to occur; (iv) rice cultivation and sea salt production will suffer serious damage in the Huanghe River (Yellow River) Delta.

Acid rain with the strongest acidity, a pH lower than 4.4, is not common in Japan. In Yokkaichi City, Mie Prefecture, Central Japan, it was shown that the effect of SO_2 was not significant, but SO_3 has a close relationship to rice yield. The relationship between rice yield and the activity of the petrochemical complex showed a linear function. In some seriously affected cities in China and Korea, the pH of rain shows strong acidity, such as 3.1 - 3.3.

The fluctuation of paddy rice yields in Japan was considered for three periods within a 100 years' trend. The second half terms of each period show strikingly greater fluctuations than the first half. The fluctuation of yields is affected by climatic fluctuation, but its magnitude is affected by technological development and socio-economic changes in agriculture. Such factors should be taken into consideration when estimating the effect of global warming.

Lastly, flow impacts of environmental change on agriculture, forestry and fishery were presented.

In East Asia, the natural environmental factors, particularly monsoons, typhoons, deltas, coastal plains and sea level rise are all important. Socio-economic factors such as population density, paddy rice production, industrial developments connected with air pollution and acid rain, should all become major topics for further discussion. Regional differences in climate change, as has been shown by Yoshino et al. (1981), are rather clear in East Asia. We should study this point more intensively in relation to the environmental changes mentioned above.

References

Chen, L.-x., Gao, S.-h., Zhao, Z.-c., Ren, Z.-h., and Tian, G.-s. (1990): Change of climate and its influence on the cropping system in China. Acta Met. Sinica, 4 (4), 464-474.

Dooge, J. C. I. (1990): World Climate Impact Studies Program. A paper presented at the Second World Climate Conference at Geneve in October, 1990.

FAO (1984): Land, food and people. Rome, 1-96.

Gornitz, V. et al. (1982): Global sea level trend in the past century. Science, 215, 1611-1614.

Huang, M.-y. et al. (1988): Acidity of cloud water and rain water in Chongqing region and its chemical composition analysis. Scientia Atmosph. Sinica, 12 (4), 390-395. (in Chinese)

Japan Meteorological Agency (1990): Climatic change in association with increasing gases with greenhouse effects (II). Investigation Working Group for Greenhouse Effects, Office of Climate Program, 47p. (in Japanese)

Jones, P. D. (1989): Global temperature variations since 1861. "A paper presented at the Symposium on the long term variability of pelagic fish populations and their environment" 14-17 November, 1989, Sendai, Japan.

Long, Siyu et al. (1989): Agroclimatic resources and its development and application in Jiangsu Province. In: Jiangsu Resources and Environment. Jiangsu Education Publ., 32-38. (in Chinese)

Manabe, S. et al. (1988): Ocean circulation and climate. Activities - FY 88, Plans - FY 89, Geophys. Fluid Dynamics Lab./ NOAA, 22-26.

Okada, M. (1988): Long term fluctuation of mean tide level. The 37th Meeting of Office of Climate Program, JMA, 41-71. (in Japanese)

Parry, M. L., Mendzhulin, G. V. and Sinha, S. (1990): The potential impact of climatic change on agriculture and forestry. Report of Working Group 2, IPCC, Chapter 2 (to be printed).

Ren, Meie (1990): Effect of sea level rise and land subsidence on the Huanghe River Delta. Scientia Geographica Sinica, 10 (1), 48-57. (in Chinese with English abstr.)

Shen, Z.-1. et al. (1989): Acidity of cloud water and rain water in Shanghai region and chemical composition analysis. Scientia Atmosph. Sinica, 13 (4), 460-466. (in Chinese)

Somboon, J. R. P. (1990): Coastal geomorphic response to future sea level rise and its implication for the low-lying areas of Bangkok metropolis. Southeast Asian Studies, 28 (2), 154-170.

Taniyama, T. (1990): Global spread of acid rain and its effects to ecosystem. Hydrology (Japan), 20 (1), 29-41. (in Japanese with English abstr.)

Uchijima, T. (1988): The effects on altitudinal shift of rice yield and cultivable area in Northern Japan. In: The Impact of Climatic Variations on Agriculture, Vol. 1: Assessment in Cool Temperate and Cold Regions, Ed. by M. Parry et al., Kluwer Acad. Publ., Dordrecht, 794-808.

Uchijima, Z. (1981): Yield variability of crops in Japan. GeoJournal, 5 (2), 151-163.

Xu, Qun et al. (1989): Tendency of climatic change; Its cause and countermeasure. In: Jiangsu Resources and Environment. Jiangsu Education Publ., 15-22. (in Chinese)

Yamamoto, R. (1990): Review of the observational studies of recent global warming. Tenki, 37 (5), 289-305. (in Japanese)

Yoshino, M. and Urushibara, K. (1981): Regionality of climatic change in Monsoon Asia. GeoJournal, 5 (2), 123-132.

Yoshino, M. (ed.) (1988 a): Problems in climatic impacts and application studies in Japan. Kisho-kenkyu noto, (160), 1-231. (in Japanese)

Yoshino, M. (ed.) (1988 b): Climatic change in the coming half century and an estimation of food, energy and water in Japan. Kikogaku-Kishogaku Kenkyu-hokoku, (14), 1-108. (in Japanese)

Yoshino, M. (1989): Network activities of Japan on climate impact studies. A paper presented at the Networkshop at NCAR, USA, on March 14, 1989. 15p.

Yoshino, M. (1990 a): Recent studies on climate impact and application; Its organizations, problems and results. Tenki, 37 (1), 5-17. (in Japanese)

Yoshino, M. (1990 b): Some problems in agricultural climatology related to the recent global warming. Nogyo oyobi engei (Agriculture and Horticulture, Tokyo), 65 (7), 417-425. (in Japanese)

Yoshino, M. (ed.) (1990 c): Global climate in the coming half century and an estimate of agriculture, fishery, forestry and human environment of Japan (I). Kikogaku-Kishogaku Kenkyu-hokoku, (15), 1-84. (in Japanese)

Zhang, Jia-cheng (1989): The CO_2 problem in climate and dryness in North China. Meteorological Monthly, 15 (3), 3-8. (in Chinese)

Zhao, Z.-c. (1989): Climatic changes in China induced by doubled CO_2 as simulated by five GCMs. Meteorological Monthly, 15 (3), 10-14. (in Chinese)

Zhu, Q.-m et al. (1989): A preliminary analysis on the effect of sea level rise on the Chanjiang delta and on the coast of Jiangsu Province. In: Jiangsu Resources and Environment. Jiangsu Education Publ., 7-14. (in Chinese)

The Global Climate Protection Strategy of the Enquete-Commission of the German Parliament

*W. Bach**

Center for Applied Climatology and Environmental Studies,
Climate and Energy Research Unit, Department of Geography,
University of Münster, Robert-Koch-Str. 26, W-4400 Münster, FRG
*Member of the Enquete-Commission

1. Climate change as part of the complex environmental problems

The climate problem must be viewed within the context of the accelerating destruction of our natural environment. Industrial and agrarian society has, by wastefully exploiting natural resources and thoughtlessly applying technology, attained a potential for inflicting damage on a scale that is seriously endangering the sustained survival of man. The chemical catastrophes of Seveso, Bhopal and Basle, the nuclear reactor catastrophes of Three Mile Island and Chernobyl, and the recurrent famines in the developing countries are early warning signs that have briefly shocked society. But no less alarming are the persevering threats, such as air, water and soil pollution, the contamination of food, the destruction of forests and the ozone layer, the depletion of genetic resources and gene manipulation, the long-term storage of radioactive and chemical wastes, the stockpiling and proliferation of atomic, chemical and biological weapons, the degradation of landscapes, rapid population growth, the growing scarcity of food, and last but not least the epidemic spread of allergies, viral infections and cancer (Fig. 1).

For a long time, nature was able to absorb this ruthless pillaging of resources and the destruction of the environment. But once a certain "threshold" has been exceeded - and that could very soon be the case - it will respond in a way that must have catastrophic

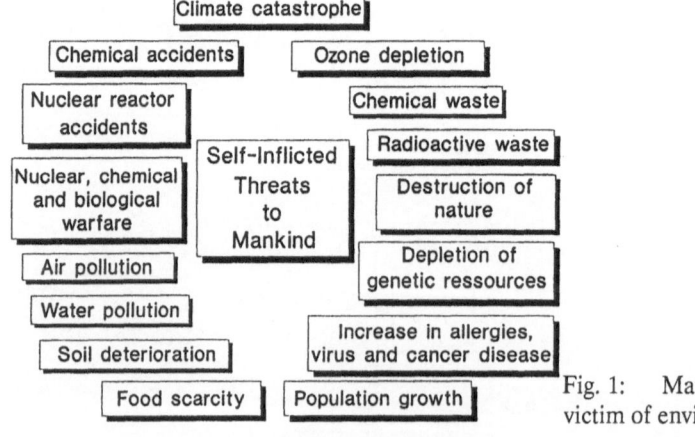

Fig. 1:   Mankind as instigator and victim of environmental hazards

The Global Environment
Editors: K. Takeuchi · M. Yoshino   © Springer-Verlag Berlin, Heidelberg 1991

consequences for man. For the world-wide threat of ozone depletion and climate change, there is still a faint glimmer of hope, because the influencing factors are known, and it appears that the willingness exists (in the case of ozone) or seems to grow (in the case of climate) to implement countermeasures. If joint action is soon taken, then we can perhaps utilize what little time is left to just barely ward off the global climate threat that is being posed to society by the additional man-made greenhouse effect with all of its far-reaching regional and local repercussions. The final decade of this century is our last chance to abandon the self-destructive logic of the present economic system. A system that is environmentally, climatically and socially benign and that takes into account the aspirations of future generations, requires a complete re-structuring of the economic and political institutions.

In the following, the climate-influencing factors and their possible impacts are described. The climate strategy developed by the Enquete-Commission of the German Parliament is summarized. The recommended measures are evaluated in terms of their effectiveness for mitigating the additional man-made greenhouse effect.

2. Climate-influencing factors

The most important climate-influencing factors are:

- The extremely wasteful use of fossil fuels in the industrialized nations, leading to greenhouse gas emissions from coal, oil and gas which are too large and too fast to be absorbed by the natural system.

- The manufacture and release of artificial products such as CFCs and halons, which are not only effective greenhouse gases, but whose waste products are also depleting the stratospheric ozone layer.

- The destruction of forests, other vegetation, and soils.

- The highly mechanized agro-business.

- The release of gases from waste dumps and the emissions from cement production, etc.

Table 1 shows that currently ca. 46 % of the greenhouse gases (37 % CO_2, 9 % CH_4 and O_3) are emitted by the energy and transportation sectors. Roughly 24 % is accounted for by CFCs and halons produced by the chemical industry. Another 18 % results from the destruction of tropical forests (12 % CO_2, 6 % CH_4, N_2O, CO etc.). Agriculture contributes around 9 %, for the most part in the form of CH_4 (from rice paddy fields and cattle feedlots) and N_2O (from artificial fertilizers). The remaining 3 % is accounted for by emissions from waste dumps and cement production. As Table 2 shows, in the past, CO_2 was the most prominent greenhouse gas with a share of ca. 60 %. During the last

Table 1 Worldwide contribution of the various sources to the additional greenhouse effect in the 80's.

Sector	Share of trace gases	Sources
Energy and Transport	46% total 37% CO_2 9% CH_4 and O_3 (O_3 formed by precursors NO_x, CO and NMVOC)	Emissions from coal, oil and gas; in the conversion sector from electricity and district heat as well as refineries; in the end-use energy sectors residential, commercial, industrial and transport
Chemical products	24% CFCs, halons etc.	Emissions from CFCs, halons, FCs
Tropical deforestation	18% total 12% CO_2 6% other trace gases such as CH_4, N_2O, CO	Emissions from burning and decomposing tropical forests, and from soils
Agriculture	9% above all CH_4 and N_2O	Emissions from rice paddy fields and ruminants (CH_4) and artificial fertilizer (N_2O)
Others	3%	Emissions from cement production and waste dumps

After: UNEP/WMO (1990) and EK (1990)

decade, emissions of CFCs have rapidly grown to take second place. The Montreal Protocol and the agreement of London should reduce the influence of the CFC-emissions on climate and the ozone layer during the decades to come.

Table 2   Relative contribution of the global concentrations
           of greenhouse gases to the additional greenhouse
           effect.

Greenhouse gases	Cumulated from 1860 - 1980[1] %	Actual values for 1980 - 1990 %
Carbon dioxide (CO_2)	60	46
Methane (CH_4)	14	9
Ozone in the troposphere (O_3)[2]	10	14
Chlorofluorocarbons (CFCs)	9	24
Nitrous oxide (N_2O)	3	4
Water vapor in the stratosphere (H_2O)	4	3

[1] Calculated with the Muenster Climate Model.

[2] Refers to the lower atmosphere of the Northern Hemisphere;
data are not derived from a complete climate/chemistry model
and are therefore uncertain.

After: EK (1990)

3. Climatic Impacts

The observed global warming is due to natural processes and to the increasing
concentrations of greenhouse gases in the atmosphere as a result of man's activities.
Because of the large heat capacity of the oceans, however, the warming effect is delayed
by several decades, so that statistical proof of the additional greenhouse effect induced by
human activities can only be obtained after a certain time lag. Since 1860 the mean global
surface temperature has risen by about 0.6 °C. If the current trend continues, the global
warming could exceed the critical rate of 0.1 °C per decade by a factor of between 3 and
5. There is reason for concern that the earth's ecosystems, having already been placed
under considerable stress by pollution, especially tropospheric ozone, acid rain, and heavy
metals, etc., will be unable to cope with this additional climatic stress. The vegetation
zones are not likely to shift polewards because of the rapidity of the changes, and
consequently there is a high probability that vegetation which is adapted to present site
conditions will die out. As the indigenous vegetation disappears, new plant forms that are
better-adapted to the changed conditions may eventually take their place.

With increased warming there will be more drought, floods, and subtropical storms. The warming will cause not only the snow and the sea ice to shrink and to become thinner, but also the Arctic permafrost soils to thaw. This will reduce the earth's reflection, enhance solar absorption, release great quantities of CH_4 and CO_2, thereby fuelling the warming process as a consequence of these feedback processes.

The sea level has already risen by about 15 cm over the course of the last century. If the present warming trend continues, an additional sea level rise on the order of 30 to 110 cm can be expected. It is almost certain that during the next few decades some 200,000 Maledivians and a few hundred thousand people living on atolls in the Pacific will have to be resettled. At a sea level rise of only 0.5 m millions of ecological refugees from Bangladesh, Egypt and other threatened, densely populated coastal regions are to be expected.

At the present rate of world population growth, the anticipated impact on the agricultural sector needs to be taken especially seriously. Agricultural yield will be reduced in vulnerable regions. World market prices for agricultural products will rise. This will result in more famines and conflicts over the distribution of food. The climatic changes triggered by human activities will thus have a destabilizing effect on world peace.

In order to ward off this threat, it is imperative to act immediately to reduce emissions of greenhouse gases. For this purpose, the Enquete-Commission of the German Parliament proposes the following climate strategy.

4. Recommendations for international and EC-wide action to reduce the additional greenhouse effect and to protect the Earth's atmosphere

In order to protect the earth's atmosphere, binding international agreements are called for. At the same time, the different types of problems regarding ozone-depleting substances, energy and climate, and protection of the tropical forests require very specific strategies for action in each individual case. Consequently, the Commission believes that the quickest way of arriving at international agreements is to take a sectoral approach. It is also of the opinion, however, that the framework for an overall protection strategy of the earth's atmosphere must be set up right at the outset to ensure a sustained ecological and economic development.

4.1 Measures and reduction targets with respect to CFCs

A total stop to production and use of CFCs and halons is not only the most effective means of protecting the ozone layer, but it is also the most economic short-term measure for reducing the additional anthropogenic greenhouse effect. In order to ensure comprehensive protection of the Earth's atmosphere, therefore, the Commission regards it as essential for the stipulations of the Montreal Protocol to be strengthened even further than was done by the agreements of the 2nd conference of the signatory states in London. Specifically, this means:

- To move up from 2010/2015 to 1997 and to completely stop production and use of all fully halogenated CFCs and halons as well as carbon tetrachloride and methyl chloroform;
- to stop by 2005 at the latest production and use of partially halogenated chlorofluorocarbons, which had not yet been included in the Montreal Protocol;
- to permit a time-limited and restricted use of radiatively-active fluorocarbons only if it can be proven that no substitutes are available;
- to eliminate all currently existing exemptions for the industrialized and developing countries; for the latter substitutes and technologies must be made available and financial assistance must be provided;
- to supply funds derived from a tax of 10 deutschmarks on each kilogram of CFCs and halons produced and 5 deutschmarks on each kilogram of H-CFCs;
- to identify worldwide all raw materials, intermediate and finished products that are made of ozone-depleting and climate-influencing substances and that should be subject to an import ban under the provisions of the Montreal Protocol;
- to provide government monitoring on production and use as well as the achieved reduction rates in a manner that is transparent to the public.

4.2   Measures for the protection of tropical forests

For the preservation of the tropical forests, the Commission has developed the following three-stage plan:

- Firstly, reducing the rate of forest destruction between 1990 and 2000 to that of 1980;
- secondly, stopping the destruction of tropical forests by 2010 at the latest, so that the total forested area in absolute terms no longer declines;
- thirdly, between 2010 and 2030 restoring the forest stand to the level of 1990.

This plan is to be financed by establishing an International Trust Fund of the order of 10 billion DM to be paid annually for the most part by the industrialized nations.

The Commission proposes that an International Convention for the Protection of Tropical Forests be adopted at the 1992 UN Conference on Enviroment and

Development in Brazil separately from an International Convention on Climate and Energy. This is reasonable, because the global climate is just one aspect of many and because here the focus should be mainly on conservation of species diversity, protection of indigenous peoples, social, ecological and economic issues, and on factors that affect the climate more on local and regional scales.

4.3 Measures and reduction targets in the areas of energy and climate

The reduction of the additional anthropogenically-induced greenhouse effect requires drastic changes in the global energy systems. These changes must be implemented as soon as possible on the basis of carefully worked-out long-term strategies. The magnitude of the reductions depends on the reduction potential that can be mobilized in each individual country. The allocation of the reduction shares must take into account both the different economic and technological situation in the various countries and criteria that ensure equitable treatment of all. It is clear that the developing countries need to catch up. A binding global plan for reducing emissions can only succeed if fair reduction shares are allocated based on a catalog of criteria that take into consideration all important aspects.

The Commission has formulated the following criteria for identifying reduction targets for individual countries:

- Economic strength
- The existing energy supply structure and the efficiency of the energy system
- Total accumulated emissions and the current per-capita emission rate
- Population development and refugee flows
- The export/import balance of energy-intensive products and processes
- Emission per unit land area and the influence of the climate on energy use (heating, refrigeration)
- The anticipated extent of climate impacts and the level of environmental awareness.

Based on present knowledge, the Commission considers the following region-specific reduction targets for energy-related CO_2 emissions, using 1987 as the base date, both necessary and feasible:

- For the Federal Republic of Germany and other economically-strong western industrialized countries with very high per-capita CO_2 emissions, reductions of at least:

- 30 % by 2005,
- 50 % by 2020, and
- 80 % by 2050.

- For economically-weaker countries in Western Europe and overseas, reductions of at least:
  - 20 % by 2005,
  - 40 % by 2020, and
  - 80 % by 2050.

- For Eastern Europe Countries including the USSR initally an even slower procedure is indicated for economical and technological reasons, leading to reductions of:
  - 10 % by 2005,
  - 30 % by 2020, and
  - 80 % by 2050.

- Equity considerations lead to a greater share in global CO_2-emissions in the developing countries as compared to the industrialized countries in the first half of the 21st century; at the same time the rate of increase of fossil fuel use per year must be slowed down so that the transition to a more efficient energy economy based on renewables is not impeded, this results in an increase of:
  - 50 % by 2005,
  - 60 % by 2020, and
  - 70 % by 2050.

- For industrialized countries taken together reductions of at least:
  - 20 % by 2005,
  - 40 % by 2020, and
  - 80 % by 2050.

- For the world as a whole, reductions of at least:
  - 5 % by 2005,
  - 20 % by 2020, and
  - 50 % by 2050.

The target formulated at the 1988 Toronto conference "The Changing Atmosphere: Implications for Global Security," namely a 20 % reduction in global CO_2 by the year

2005, is, as the foregoing discussion shows, not practicable. The reason is that the economically-strong industrialized countries would be required to implement reductions of considerably more than 30 % in order to offset the increases in the developing countries.

In order to mobilize the existing emission reduction potential, the Commission proposes that, together with a Reduction Protocol an International Trust Fund be established, initially having DM 20 billion at its disposal. In addition, an International Environment Council should be set up under the auspices of the United Nations for the purposes of strengthening global environmental policies and supervising the measures to be implemented.

4.4 Overall strategy for the protection of the Earth's atmosphere

The different sectoral and international agreements must be integrated into an overall strategy. The Commission recommends that the timetable shown in Table 3 be adopted, culminating in the ratification of an overall strategy on the protection of the Earth's atmosphere.

Because of the lack of reliable estimates on the emission reduction potential, it is not yet possible at this time to establish concrete reduction targets for individual countries. In the next section, the Federal Republic of Germany is taken as an example to demonstrate what detailed studies are needed and which measures can be applied to attain the required reduction goals.

5. Recommended action for reducing energy-related climatically-relevant trace gas emissions in the Federal Republic of Germany

In order to prevent further aggravation of the additional anthropogenic greenhouse effect and in the light of what is already known, the Commission considers it essential to take appropriate steps without delay to reduce emissions in the Federal Republic of Germany. Besides a coordinated approach in the EC, in the opinion of the Commission it is also urgently necessary for the Parliament and the Federal Government to initiate action at the national level that will permit the Federal Republic of Germany to assume a leading role. It is imperative to act now and not to wait until all the respective international agreements have been made.

Due to lack of time, so far the Commission has primarily concerned itself with energy-related climatically-relevant trace gas emissions. In the next study phase, attention will focus on such interrelated topics as "ecosystems - climate - agriculture", which have an enormous importance for food security.

Table 3   The 1990s, a Decade of Critical Decisions

Time-table for the development of an overall strategy
for the protection of the Earth's atmosphere as prepared
by the Enquete-Commission of the German Parliament

1990   Recommendation by the Second World Climate Conference in
Geneva to negotiate international conventions on
o   Climate and Energy and
o   Protection of Tropical Rainforests
to be completed at the UN Conference on Enviroment and
Development (UNCED) in 1992 in Brasil.

1991   Two preparation conferences held to negotiate the draft
conventions.

1992   Acceptance of the two international conventions in
Brasil and strengthening of the Montreal Protocol
according to the version agreed upon at the conference
of contracting nations in London.

1994   Acceptance of concrete implementation protocols to both
conventions.

1995   Ratification and implementation of the protocols and the
international funding mechanism.

1998   Preparatory Conference for combining the sectoral
international conventions.

2000   Acceptance of a complete strategy for the protection of
the Earth's atmosphere.

After: EK (1990)

5.1 Reduction targets for energy-related climatically-relevant trace gas emissions

An initial evaluation of the potential for reducing emissions, carried out for the Commission by some 50 different institutes and laid down in a total of around 150 energy studies, has yielded the reduction plan shown in Table 4 for CO_2 and other directly or indirectly climatically relevant trace gases. Because of the different energy structure in the eastern part of Germany, altered reduction contributions could result in some of the sectors as a consequence of the former GDR joining the Federal Republic of Germany. Taken as a whole, however, as already shown for the case of CO_2, the recommended reduction targets (expressed as percentages) will not be changed because of German unity.

5.2 Assessment of the CO_2 emission reduction potential

Within the scope of the Commission's energy study program the CO_2 emission reduction potential for the year 2005 as compared to the reference year 1987 was estimated for the following three scenarios:

- A reduction scenario "energy policy" in which the share of electricity from nuclear energy of 31.2 % in 1987 is maintained.

- A reduction scenario " nuclear energy phase - out".

- A reduction scenario "nuclear energy expansion" (in which the share of electricity from nuclear reactors is almost doubled to 60 % and the share in primary energy consumption is more than doubled from 11.4 % in 1987 to 24.6 % in 2005).

The most significant result shown in Table 5 is that a considerable increase in the use of atomic energy would not cut down on CO_2 emissions as compared to a phase-out of nuclear power. In all three scenarios the CO_2 emission reduction potential is almost the same, at 34-36 %. There are, however, appreciable differences among the various sectors, with the greatest emission reductions being achieved in the end-use energy sector for a nuclear phase-out, while the energy conversion sector would account for the largest reductions with a nuclear energy expansion. However, these reductions are not independent of one another, since energy savings in the end-use energy sector (particularly with respect to electricity) would also lead to reduced energy consumption and emissions in the conversion sector. It is also conspicuous that, in the end-use energy sector, the greatest potential for reduction exists in private households and - to a lesser extent - in the category "commercial". An enormous, still virtually untapped savings potential exists here in connection with electricity and heating, and thus measures to reduce all forms of electricity use would be particularly worthwhile. By contrast, only smaller reductions or even increases in emissions are obtained in the sectors of industry and transportation. This could be partially attributable to the rate of economic growth of

Table 4 Reduction plan for the energy-related and climate-relevant trace gas emissions of the F.R. Germany plus the GDR for 2005 (binding reduction target) and the years 2020 and 2050 (goals) related to base year 1987.

Trace gas	Base year 1987 Mio t	Reduction target by 2005 %	Goals by 2020 %	Goals by 2050 %
Carbon dioxide (CO_2)				
FRG	715	-30	-50	-80
FRG + GDR	1067	-30	-50	-80
Methane (CH_4)	1.8	-30	-50	-80
Nitrogen oxides (NO_x)	2.6	-50	-70	-90
Carbon monoxide (CO)	8	-60	-75	-90
Non-methane volatile organic compounds (NMVOC)	1.5	-80	-90	-95

After: EK (1990)

2.4 % per annum which was assumed in all scenarios, as well as to significant increases in traffic. If appropriate political action were taken, it would be possible to mobilize a considerable reduction potential in the industry and transportation sectors as well. Of importance is the fact that in the end-use energy sector an expansion of nuclear power would lead to an emission reduction of only 9.8 % as compared to 25 % if nuclear power were phased-out, thereby hindering rather than facilitating the mobilization of the CO_2 reduction potential.

Table 6 shows the changes in use of the various energy carriers on which the scenarios were based. Compared to nuclear phase-out, with nuclear expansion more hard and brown coal but less mineral oil would be used. Substitution of the other energy carriers by the more CO_2-benign natural gas is nearly three times as high in the phase-out scenario as in the expansion scenario. Expansion of nuclear power would tend to inhibit rather than to promote the use of renewable energy sources. In total, the phase-out of nuclear power would lead to more than twice as much energy saved as compared to expansion; and, moreover, less would be consumed than in the "energy policy" reduction scenario.

Additionally, the Commission requested the "Öko-Institut" to assess the consequences of a nuclear phase-out as early as 1995. Interestingly, this scenario revealed

Table 5 Reduction potential of CO_2-emissions[1] by sector and scenario for the F.R. Germany in 2005 as related to 1987

Sector	Reduction potential (%) until 2005 for scenarios		
	Energy policy	Phase-out of nuclear energy	Expansion of nuclear energy
Residential	-8.5	-10.1	-7.1
Commercial	-3.3	-5.0	-2.8
Industry[2]	-0.6	-2.5	+1.8
Transport	-1.9	-3.2	+0.1
Sectors not considered	+0.4	-0.1	-0.1
Substitution by cogeneration	-2.8	-4.1	-1.7
Sum of end-use energy sector	-16.8	-25.0	-9.8
Cogeneration and heating plants	+3.6	+5.2	-4.1
Other electricity production	-13.3	-6.9	-15.0
Other contributions and statistical differences	-2.2	-2.0	-1.9
Sum of conversion sector	-11.9	-3.7	-21.0
Sum of end-use energy and conversion sectors	-28.7	-28.7	-30.8
Energy-conscious behavior	-5.0	-5.0	-5.0
Total reduction	-33.7	-33.7	-35.8

[1] Referring to end-use energy including the German share in international air traffic but excluding non-energy use.
[2] Including fossil fuel emissions from internally produced and used electricity.

After: EK (1990)

Table 6   Change in energy use (%) by energy carrier[1] in the reduction scenarios for the F.R. Germany in 2005 as related to 1987.

Energy carrier	Change (%) in the scenarios until 2005		
	Energy policy	Phase-out of nuclear energy	Expansion of nuclear energy
Hard coal	-6.7	-4.4	-7.3
Lignite	-3.7	-4.3	-5.2
Mineral oil	-13.9	-20.4	-12.3
Natural gas	+4.2	+10.0	+3.4
Sum of fossil fuels	-20.1	-19.1	-21.4
Hydrogen	+0.3	+0.3	+0.3
Wind energy	+0.4	+0.8	+0.1
Photovoltaics	0	+0.1	0
Renewables in the end-use energy sector	+3.6	+5.6	+1.7
Net electricity import	0	0	0
Nuclear energy	+1.8	- 11.4	+11.1
Sum	-13.9	- 23.7	-8.2
Energy-conscious behavior	-5.0	- 5.0	-5.0
Total reduction	-18.9	- 28.7	-13.2

[1] Deviations from primary energy use because non-energy use and other energy trade etc. are not taken into consideration.

After: EK(1990)

that, compared to the "energy policy" scenario, the accumulated CO_2 emissions from 1987 to 2005 would only be 1.7 % higher in the "1995 nuclear phase-out" scenario. This is due to the greater mobilization of the efficiency potential and the greater utilization of renewable energy as well as to the assumption that ambitious measures will be implemented in the transportation sector. The reliable supply of electricity is also ensured. In addition, the "1995 nuclear phase-out" scenario would not only reduce the risk of nuclear accidents to zero within just 5 years, but would also minimize both the danger of proliferation of atomic weapons and the amount of nuclear waste.

5.3 Strategy for attaining the reduction targets

In order to achieve emission reductions in Germany of 30 % each for CO_2 and CH_4, 50 % for NO_x, 60 % for CO, and 80 % for non-methane volatile organic compounds by the year 2005 as compared to 1987, the Commission recommends that

- the Federal Government initiate a crash program of far-reaching measures on the basis of existing legislation;

- the Federal Government submit the drafts of additional amendments by 1 September 1991;

- the Federal Government submit a report on the required legislative amendments to the German Bundestag by 1 December 1991, elaborate a plan for the financial policy measures and present the corresponding draft bills by 1 July 1992;

- the Federal Government submit progress reports on the required environmental, energy, and transportation legislation to Parliament every 2 years, beginning on 1 December 1993.

Within this process of dynamic adjustment and reassessment, the focus is as follows:

- **In the energy sector:**

  The Electric Utilities Act, the Energy Conservation Act, the Thermal Insulation Regulation, the Federal Tariff Code, the Waste Heat Recovery Ordinance, the Heating Systems Ordinance, the 3rd Electricity Generation Act, the Federal Pollution Control Act, the Low-Income Housing Act, etc. These must be supplemented by appropriate new legislation, e.g. regulation of refunds on electricity sold to utilities as well as energy prices that will encourage more efficient use.

- **In the transportation sector:**

  Integrated and environmentally-benign transportation systems are required shifting transportation services to environmentally compatible types of transport, minimizing increases in road traffic, improving transportation efficiency, and moderating traffic loads. In addition - among other things - the Right of Way Act and the Urban Transportation Funding Act must be changed.

5.4 Priorities for the reduction measures

In order to reach the set reduction targets, the following basic principles must be observed:

- Energy conservation by more efficient use and conversion, as well as by more energy-conscious behavior, must be given top priority.

- Renewable energy use must be extended as a matter of greatest urgency so that it will account for a growing proportion, and ultimately the largest proportion, of the energy supply.

- During the transition period toward a low-pollution energy economy, fossil fuels (coal, oil and gas) must be used as environmentally and climatically benign as possible.

- The views vary widely on whether nuclear power should be used in the future. The serious problems involved should be further discussed and the justification of continued use should be clarified.

If the goal is to prevent continued increases in energy use and emissions in spite of future economic growth and instead achieve substantial decreases, then full advantage must be taken of the technological possibilities. Space does not permit a detailed discussion of the technological potential for reducing CO_2 emissions by means of more efficient energy use as determined by the energy studies carried out for the Commission;

Table 7   Current (1987) technical CO_2-reduction potential through rational energy use in the FR Germany

Sector	Energy use PJ	Reduction potential %
Old housing stock	1600	70-90
New housing	300[1]	70-80
Commercial	1290	40-70
Electric appliances	270	30-70
Cars and aircraft	1420	50-60
Warm water use	200	10-50
Buses, trucks, fuels in industry, powerplants and refineries	6400	15-25
Cogeneration and electricity in industry	950	10-15
Total	11350	35-45

[1] in 2005

After: EK (1990)

however, a brief overview is given in Table 7. The potentially achievable reductions range from about 10 % through cogeneration to approx. 90 % by upgrading the existing housing stock.

Finally, it is shown how effective the global reduction measures recommended by the Commission will be.

6. Effectiveness of the measures recommended by the Enquete-Commission for mitigating the additional man-made greenhouse effect

The Commission has developed four scenarios on CFC reduction and three scenarios on the reduction of CO_2 emissions. Due to the tight time schedule and the precarious data situation, the development of reduction scenarios for CH_4 and N_2O was postponed for the time being. The scenarios include 14 of the most important trace gases contributing to the additional anthropogenic greenhouse effect, namely CO_2, CH_4, N_2O, CFCs 11, 12, 113, 114 and 115, H-CFC 22, halon 1301, carbon tetrachloride, methyl chloroform, tropospheric ozone, and stratospheric water vapor. The time frame extends from the preindustrial conditions in 1860 to the year 2100. The scenario results are expressed relative to 1860 in order to include the climatic burden of past years. An average rate of temperature change of 0.1 °C per decade critical for many ecosystems, was used here for a preliminary evaluation of the results. This rate of change adds up to a mean global warming ceiling of about 2 °C in 2100 as compared to the preindustrial level. All calculations were performed with the one dimensional Muenster Climate Model. As with all replicas of reality, the results from this modeling effort have a certain degree of uncertainty. In the following this so-called climate sensitivity of the model calculations is taken into account by giving the average values, as well as (in brackets) the uncertainty range.

6.1 Results of the scenario calculations

The scenario analyses allow the following conclusions to be drawn on the effectiveness of different reduction measures (see Figure 2).

- **Without measures**

    If the current trend continues (business-as-usual), then a mean global warming of 2.95 °C (2.03 to 4.19) is to be expected by the year 2100, thus exceeding the upper warming ceiling by more than 100 %.

- **Strengthening of the Montreal Protocol in the version adopted by the 2nd conference of its signatory states in London**

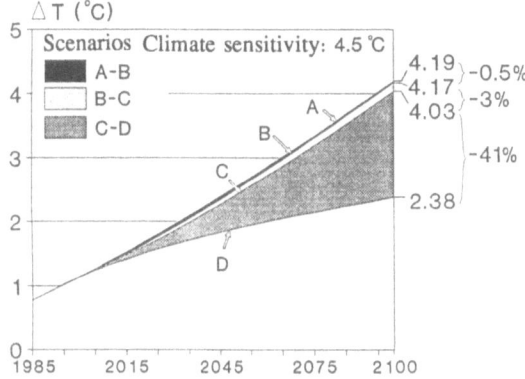

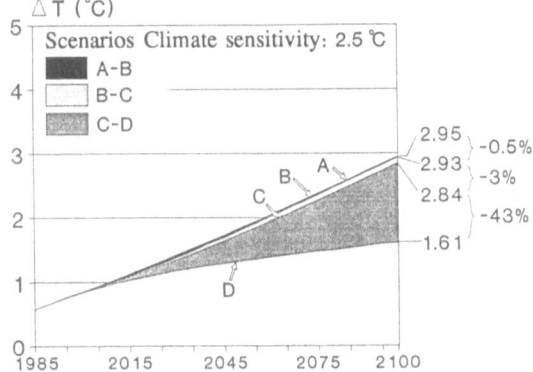

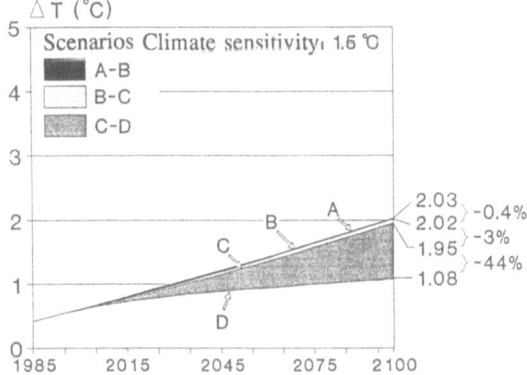

Fig. 2: Effectiveness of the measures of the Enquete-Commission to reduce the additional greenhouse effect. A-B: Strengthening of the Montreal Protocol as revised by the second conference of contracting nations in London; B-C: Tropical forest conservation plan; C-D: CO_2-emission reduction targets for fossil fuel use. The temperature change as related to the preindustrial level of 1860 was calculated with the Muenster Climate Model.

Source: Bach a. Jain

The Montreal Protocol in the London version with exemptions will phase-out the production of CFCs and methyl chloroform as late as 2010 and 2015, respectively, and it does not contain any provisions for partially-halogenated substitutes such as H-CFC 22. The recommendations of the Commission would, if adopted, remove all exemptions and lead to a complete production stop by 1995 in the Federal Republic of Germany, by 1997 for the rest of the EC, and by 2000 for the entire world. H-CFC 22 would be phased-out by 2005.

Conclusion: The strengthening of the current Protocol in its London version results only in a small reduction of the global warming from 2.95 °C (2.03 to 4.19) to 2.93 °C (2.02 to 4.17). The strengthening of the regulations is nevertheless important and necessary in order to protect the stratospheric ozone layer. The upper warming ceiling of 2 °C is still exceeded even at the lowest climate sensitivity.

- **Plan to save the tropical forests**

The three-phase plan developed by the Commission is aimed at lowering the deforestation rate in the tropics to below the level of 1980 by the year 2000 in the first phase, to halt the destruction of tropical forests completely by 2010 at the latest in the second phase, and in the third phase (2010 to 2030) to restore the tropical forest stand to the level of 1990.

Conclusion: Through these measures global warming will be reduced from 2.93 °C (2.02 to 4.17) to 2.84 °C (1.95 to 4.03). As a consequence, the temperature rise will drop to slightly below the 2 °C ceiling at the low climate sensitivity. This calculation only takes CO_2 into account, and not the many other greenhouse gases that are also emitted when forests are burned. Of importance is the fact that the burning of a primary forest releases enormous amounts of CO_2, while replanting a secondary forest on the same land area is only capable, over a period of many decades, of sequestering only about half of the originally released CO_2 from the atmosphere and fixing it in the biomass. The preservation of tropical forests has high priority not only on ecological and climatic grounds, but also for reasons to preserve the living space for indigenous peoples.

- **Reduction targets for energy-related CO_2 emissions**

The Commission is convinced that its reduction plan of energy-related CO_2 emissions for the Federal Republic of Germany, calling for cuts of 30 % by 2005, 50 % by 2020 and 80 percent by the year 2050 with respect to 1987 levels, can also be applied to other economically strong industrialized countries with particularly high per-capita emission rates. For economically weaker industrialized countries, a slower reduction rate, namely 20 % by 2005, 40 percent by 2020, and 80 % by 2050 is proposed. The industrialized countries of Eastern Europe, including the USSR, for technical and financial reasons initially can be expected to follow smaller reduction rates, namely 10 % by 2005 and 30 % by 2020, while by 2050 a reduction of 80 % should then also be attainable. Equity considerations lead to a greater share in global CO_2-emissions in the developing countries as compared to the industrialized nations, namely an increase of 50 % by 2005, 60 % by 2020, and 70 % by 2050. At the same time, the rate of increase is slowed down so that the transition to a more efficient energy economy based on renewables is not impeded. For the industrialized and developing countries taken together, CO_2 emissions reductions of 5 % by 2005, 20 % by 2020 and 50 % by 2050 will result. But for operability it is clear that reduction targets for individual countries must be allocated in a differentiated manner.

Conclusion: The additional anthropogenic greenhouse effect can be substantially mitigated by energy-related CO_2 emission reductions, with the result that the degree of warming will drop from 2.84 °C (1.95 to 4.03) to 1.61 °C (1.08 to 2.38), or by ca. 40 %.

6.2   Evaluation of the results

If the high climate sensitivity should turn out to be the most realistic, the global warming ceiling of 2 °C would still be exceeded by 0.4 °C or about 20 % despite the measures recommended by the Commission. Additional action would be necessary. This could be directed at reducing emissions of CH_4 and N_2O. In the above scenarios, from 1985 to 2100 the CH_4 and N_2O emissions rise by 96 % and 40 %, respectively. Additional model calculations could reveal which combinations of CH_4 and N_2O reductions would be required to keep global warming below the upper limit of 2 °C.

The present results give a first indication of the priorities of the measures that have to be taken.

- If steps are taken to ensure that partially halogenated H-CFCs, as is assumed here, are no longer in use by the year 2040, and if production does not increase beyond the projected levels, then additional agreements going beyond the stipulations of the Montreal Protocol as formulated in the version of the 2nd conference of the signatory states in London will only have a small additional impact on global warming. They are nonetheless important and necessary in order to protect the stratospheric ozone layer.

- The plan to save the tropical forests will make a signifikant contribution to CO_2 emission reduction by lowering the deforestation rates and by initiating systematic reforestation during the first half of the 21st century. By year 2100, the reference year in the present model calculations, the mitigating effect has, however, become relatively small. Nevertheless, preservation of the tropical forests deserves highest priority not only for ecological reasons, but also, to preserve the living space of indigenous peoples.

- The reduction of energy-related CO_2 emissions is by far the most effective way of combatting the additional man-made greenhouse effect. The greatest results can be expected in industrialized countries that have the highest technological know-how and the greatest economic strength. Transfer of technical and economic assistance to other countries could also mobilize relatively quickly the large reduction potential there.

- Further measures are necessary including those for CH_4 and N_2O.

7. Climate protection policy

The scientific facts convey a message that is unequivocal: The impact of human activities on the Earth's climate has reached a critical level. This is because of the unimpeded growth dynamic that underpins our society. For Man to survive, the suicidal logic of growth must be abondoned and our systems must be restructured in an socially, environmentally, and climatically benign way.

A comprehensive climate protection policy is therefore urgently needed. The industrialized nations must assume a leading role in implementing the necessary far-reaching reforms to restructure the present systems, since it is they who have the technological know-how and the financial power. Moreover, in their own best interest, the industrialized countries must support the developing countries in an unselfish manner.

A tractable climate protection policy requires a change in the outlook of the principal actors:

- The **energy companies** must be more open to efficient energy use, decentralized energy production, cogeneration, and an increasing utilization of renewable energy sources. They must reduce the wasteful use of energy. Moreover, they must take more notice of the energy services needed by consumers.

- The **chemical industry** must stop production of ever new substances that have unpredictable impacts on the environment and the climate.

- **Agrobusiness**, which is destroying the environment, the landscapes, rural farming and forestry, notably in the developing countries, must be replaced by ecologically-compatible forms of agriculture and forestry that preserve the environment.

- Conventional **traffic planning** is deleterious to the environment, landscapes, forests, health, and climate. To avoid a catastrophic collapse of the entire transportation system, individual motorized traffic must be speedily replaced by an integrated transportation concept that emphasizes railways and local forms of public transport together with urban planning that reduces the need for traveling.

The success of a climate protection policy will hinge on whether or not, and if so, how fast the numerous possibilities for action can be put into practice. The creativity and commitment of each and every one of us are called for. We owe it to our descendants to maintain our planet's ability to support life.

8. References

Bach, W. (1990): Contribution to Man-Nature-Technology, Expo '92, Sevilla; unpublished Manuscript.

Bach, W. und A. K. Jain (1990): The effectiveness of the measures recommended by the Enquete-Commission "Protecting the Earth's Atmosphere" of the German Parliament.

Enquete-Commission (EK) (1990): Protecting the Earth. A Status Report with recommendation for a new energy policy. Economica Bonn/C. F. Müller Verlag, Karlsruhe.

Intergovernmental Panel on Climatic Change (IPCC) (1990): Policymakers Summary. Working Group Reports 1 - III.

The Global Warming Issue – A Social Scientist's View

K. Takeuchi

Research Center for Advanced Science and Technology,
University of Tokyo, 6-1 Komaba 4-Chome, Meguro-ku, Tokyo 153, Japan

As a statistician long affiliated with the Faculty of Economics, University of Tokyo, I am able to talk here as a social scientist, concerned very much with the global warming problem. I am not, however, a specialist in any of the related scientific disciplines and I am mainly interested in the socio-political implications of global warming and policy issues which arise as a consequence.

There are many unsettled questions about the causal relationships between CO_2 emissions and the global warming issue and there exists a wide spectrum of opinions among specialists in the relevant fields. As a social scientist, and a layman in the natural sciences, I am in no position to take sides with one opinion or another where controversy arises, but I think it a part of my duty to consider how to understand the uncertainty of the problem involved, and to formulate acceptable and effective policies in the face of it.

One important aspect of the uncertainty is that it is not quantitative but qualitative. I mean this in the sense that we don't have a continuous probability distribution over a range of quantitative variables, but we have some, small or not small, probabilities for quantum jumps, which may have calamitous effects. In such cases computation of expectation has little relevance, but assessment of possible calamitous cases and estimation, inevitably rough, of the magnitude of their probabilities are important.

In a system as complex as that of the global climate the capacity to maintain the present equilibrium is limited. With small perturbation the equilibrium is restored, but if perturbation exceeds the threshold, the whole system moves away from the original equilibrium, and will settle at another point. It is also often observed that the shift from one equilibrium to another is abrupt and quick, and the cause which brings about such a shift is often unclear or apparently small. It has been recently shown that this is a common feature of various types of non-linear multi-dimensional dynamic systems and can be viewed as a kind of "chaos" in a mathematical sense. Chaos theory is still in an early stage of development, and we cannot expect much in the way of definitive results concerning specific systems. We can however derive some insight into complex dynamic systems. As I understand it, the global geophysical system has these characteristics and the fluctuations of temperature between glacial and inter-glacial periods, and also smaller scale changes in inter-glacial periods, should not be considered as smooth cyclical changes, but can better be seen as jumps of a chaotic nature. If that is the case, pressure on the global environment caused by human activities could well trigger off

The Global Environment
Editors: K. Takeuchi · M. Yoshino © Springer-Verlag Berlin, Heidelberg 1991

such a chaotic shift, and no one can tell what would be the new level of equilibrium.

Although not a physical scientist, I believe that emissions of CO_2 and other greenhouse gases could exceed the threshold point and cause a chaotic shift in the global temperature if the present trend, not the present level, continues for the next 100 years. At present both the level of atmospheric CO_2 concentrations and average global temperatures are at the highest point reached in recent history. If both continue to rise substantially, the human race will be entering a completely new climatic environment.

It must also be emphasized that any social system is highly flexible. It can be transformed to conform to changes in external or internal conditions to an unexpectedly great extent. It can also sometimes collapse and disintegrate when stresses are too strong. This means that in social systems no quantitative relationship between social factors is permanently stable. Hence, the relationships between energy consumption and GNP, or CO_2 emissions and GNP vary greatly from country to country, and even within a country from time to time. Developed countries have substantially decreased these ratios since the oil crisis of 1973, but there are still large differences even among developed countries. For developing countries and communist and former communist countries these ratios are usually high. This is partly a reflection of the differences in industrial structures in the communist countries but such a large differential as 10 to 1, which is the case between China and Japan, can only be explained by the differences in levels of technology. These figures indicate that there is considerable room to reduce CO_2 emissions substantially, without adversely affecting the economy; or to improve economic well being without increasing CO_2 emissions in the less developed countries. The often discussed trade-off between CO_2 emissions control and economic growth is not rigid, even if not altogether false, in the long run.

In this context we have to consider four different levels of quantitative relationships in a social system. The first level is that of theoretical possibility. There are basic, almost physically necessary relationships among some quantities in a social system, such as the amount of food and the size of the population. However, we don't need to limit our consideration to the bare necessities for survival. We can also consider some social and cultural needs as physical necessities. We could, for example, calculate the minimum necessary amount of energy consumption and associated CO_2 emissions to guarantee a healthy and civilized life for the whole population. It is certain though, that such a theoretical minimum would turn out to be far below the actual level.

The second level is technological feasibility. The theoretical possibility defined above cannot be realized off-hand. We must start with already existing physical systems and make use of thus far available technologies. In this way come to the concept of technological feasibility. The technologically feasible minimum level of energy consumption would be higher than the theoretically necessary level but would still be much lower than the actual level.

The third level is that of economic accountability. Technological feasibility cannot be realized if its cost is too high or requires too large an investment, and viable technologies must be feasible economically as well as technologically.

There is still a fourth level of socio-political reality which prevents even economically efficient technologies from being implemented because of such factors as the power of vested interests, social inertia and social conflicts. These account for some, and not necessarily a small part, of the gap between theoretical possibility and actuality.

I have referred to these factors not so much to emphasize the "realistic-pessimistic" view that socio-political realities are what always prevail, but rather the "realistic-optimistic" view that although each of the above four levels must be taken into account, all can be modified. We may quote several examples to show that socio-economic realities and even economic accountabilities can change if sufficiently powerful external or internal pressures are exerted.

The most obvious example is the dramatic increase in efficiency of energy usage per unit of GNP output after the 1973 oil shock in many of the developed countries. In Japan the usage in 1985 was 60% of the 1973 figure.

We should also take note of the dynamic or unstable character of contemporary society. The economic system of the advanced societies, which can be characterized as capitalist and industrial, has a strong propensity to expand, and if it is not growing at a certain rate, it results in depression and unemployment. Developing societies on the other hand, have an absolute need for economic growth, even to maintain the barest essentials for survival of their growing population. They need much more to increase their level of economic well being and to catch up with the advanced societies. In view of the fact that the difference of per capita GNP between the richest and the poorest countries amounts to the outrageous ratio of 100 to 1, and also that even the richest countries still want and need economic expansion, the total GWP (gross world product) must grow substantially if social stability and some degree of equity and justice are to be achieved.

In the 21st century the world population will surely reach 10 billion, twice the present level. If economic well being can be properly expressed in terms of GNP, its per capita GWP must increase at least 2 to 3 times. This means the total GWP must grow 4 to 6 times the present level, i.e., annual growth rates must be at least 2.8% to 3.6%, if the desired level is to be reached in 50 years.

The 20th century, especially its latter half, has been a period of exceptionally rapid world wide growth, in terms of population and production. The annual growth rate of GWP has been roughly 4%, which if continued for another 50 years will push total GWP up to 8 times the present level.

Let us assume, rather hopefully, that neither major wars nor social upheavals nor large scale natural disasters will disrupt the course of social development, and that limitation of natural resources will not occur, in the first half of the 21st century. In this case the developed countries will move to the "post-industrial" era and most of the developing countries will undergo a process of industrialization. The overall average growth rate of GWP will be 3% to 4% annually. But if development is left to take its "natural" course, world energy consumption will increase roughly proportionally to GWP and CO_2 emissions a little less than that, which means 4 to 6 times the present level of energy consumption and 3 to 4 times

CO_2 emissions.   Such levels could be really dangerous because of the warming effect.

In the light of our basic understanding of the problem, we must move towards formulating worldwide goals. Tentatively I propose this scenario, using the year 2050 as a benchmark:
1.   The world's population will be roughly 10 billion.
2.   Accepting per capita GWP as an appropriate measure of economic well being, its total level will be 5 times the present level.
3.   World total energy consumption will be twice the present level.
4.   Total CO_2 emissions will be the same as at present.

This means that per capita GWP will increase 2.5 times, energy consumption per GWP will decrease to 40% of present levels, and CO_2 emissions per unit of energy consumption will be halved.

Such a goal, though ambitious, would seem to be well within technological feasibility, and should be achievable without major technological break-throughs.

I don't wish to present the above figures as definitive, but to emphasize that we must clarify, however roughly, the target year and the total worldwide levels of the main variables before discussing policy problems.

If we accept such overall figures as a goal at a specified year, our next step is to consider the distribution of each quantity to various regions and countries.   The following are the most important implications of the figures suggested:
(a) Overall growth rates for developed countries must be substantially below the average, at least of their domestic consumption. (Although not necessary of their production.   The difference would be used for international aid or worldwide investment.)
(b) Efficiency in energy consumption and CO_2 emissions in developing countries must be greatly increased.   Even though per capita energy consumption in these countries is small, the overall contribution of all those countries in energy saving and emission reduction could be substantial.

It must also be emphasized that the rate of reduction of energy consumption or CO_2 emissions must be different from country to country, depending on their respective industrial structures and levels of technological development.

In order to achieve such a goal, we must take strong and consistent policy measures.   It is clear that market forces alone cannot take care of such long range problems as CO_2 emissions.

Any policy measures taken must be beneficial in the long run, whatever the exact quantitative causal relationship of the CO_2 greenhouse effect will turn out to be.   The abolition of the use of CFC's comes into this category provided that their substitutes have no strong greenhouse effect.   So, any measure to decrease CO_2 emissions which simultaneously increases energy efficiency is good, again provided that it does not introduce further hazards.

But in the long term nature and the uncertainties of the CO_2 global warming issue do not justify its negligence, or an attitude of "wait and see".   It will also take time to formulate proper policy measures, to get them accepted and implemented, and finally to have them produce the desired results.

The most important characteristic of any proposed policy measures would seem to be decreasing the gap between technological feasibility and socio-economic reality, rather than widening technological scope (although technological development is also necessary). And in order to reduce the gap, some aspects of the socio-economic system and human life styles must be altered. For example, the use of private automobiles could at least be partly replace by more energy saving transportation methods. But this would also entail the restructuring of cities, and changing the life style of their citizens.

It is not my intention here to propose any specific policy measures to reduce CO_2 emissions. Rather it is to emphasize the importance of the long term consistency and coherence of all economic, industrial and technological policies, in relation to energy and environment problems, and the need for the proper assessment of any policy measures which may have some effect on them.

One important aspect of the issue is the length of time involved. Fifty years is too long a period over which to formulate consistent and coherent policy plans. There will be too much change in all countries within the time span of half a century to envisage any steady, sustained and coordinated implementation of policies over 50 years. Look back and see how much the world has changed in the last 50 years, since the beginning of the Second World War! What will happen in the next 50 years? Even national boundaries may not exist 50 years from now.

Therefore, to formulate policies for the year 2050, say, is inevitably unrealistic. Shorter terms must be chosen for planning. More realistic will be to take 2010 or 2020 as target years.

The priority is to decide on the optional means for increasing efficiency of energy consumption in each country and to formulate practical plans to achieve specified goals, both within each country and also by international cooperation.

It is important to consider the problem of how to implement policy measures as well as to formulate them. Abstract consideration of such techniques as "direct control", "regulation" and the use of market forces is not enough, since policy measures can be rendered counterproductive by social reactions, if they are implemented without sufficient understanding of the relevant social or economic structures.

For any policy measure to be effective, it must be accepted as necessary by the public. Such a requirement is imperative in democracies, but it is also necessary in totalitarian states. Otherwise regulations will be evaded or distorted and will become ineffective. For the CO_2 global warming issue, therefore, the wide dissemination of factual information is most important, since the increase of CO_2 in the atmosphere cannot be smelt or felt, and even overall global warming cannot be easily identified.

It is most important to encourage a spirit of international cooperation, and to establish a worldwide understanding of the fact that in issues such as these all humanity has a common vital interest. We may think of it as a global security issue, where the survival of human civilization is at stake, in the face of the possibly malevolent power of natural forces. When a security issue is discussed, the enemy's intention is always uncertain, and we have to be prepared for all eventualities. If the U.N.'s security system to safeguard conflict between states can function at least to some extent, we can hope that a worldwide security system to guard against natural forces could be successfully introduced. We could propose the establishment of a Global Environment Security Council within the U.N., with power to intervene in the domestic policies of member states in cases where these endanger the global human environment.

The Tropical Forest Ecosystem: Reviewing the Effects of Deforestation on Climate and Environment

M. Domroes

Department of Geography, University of Mainz,
Postfach 3980, W-6500 Mainz 1, FRG

Abstract. The rapid decline of tropical rain forests in the 1980s, particularly caused by man-made forest clearance for farmland and timber exploitation, is accompanied by adverse climatic and environmental effects. In particular, rainfall and temperature conditions are adversely changed, with lower rainfall and higher temperatures. Through deforestation, the atmospheric emissions of CO_2 are increased and thus the global greenhouse effect is being strengthened. Tropical deforestation also increases soil erosion, depending upon the type of ecosystem. It also can be seen that floods more often occur in cases of deforestation. An international "action programme" for the protection and preservation of tropical forests is urgently called for as a scientific and political contribution towards the protection of the earth's atmosphere.

1. Introduction

The 1980s were a serious, but to some extent also encouraging decade with respect to tropical forests - serious, because of the persistently continuing process and extension of man-made deforestation which has even accelerated towards the end of the 1980s; encouraging, due to some first efforts by international organizations and institutions as well as national governments towards international action programmes for the sake of the preservation and protection of the remaining parts of tropical forests.

Among other efforts, particular attention should be paid to the International Tropical Forestry Action Plan (TFAP), established by the Food and Agriculture Organization (FAO) in 1986 and set up for a 5-year period until 1991. This plan has been the first internationally funded initiative designed to preserve the tropical rain forests. Besides the FAO, the UNDP as well as the World Bank and the World Resources Institute were also engaged in the Plan which served as a coordination platform for international aid aimed at the preservation of tropical forests. Noteworthy action has also been initiated by the International Tropical Timber Agreement (ITTA), adopted in 1987, in the framework of the UNCTAD Commodity Agreement. The main purpose is to promote timber trade and wood processing in the producer countries, with accompanying measures designed to protect the tropical forests and to ensure sustainable timber production. Recently, in some cases, the very opposite effects were reported as a result of encouraging forest cutting instead of the protection

The Global Environment
Editors: K. Takeuchi · M. Yoshino   © Springer-Verlag Berlin, Heidelberg 1991

and preservation of forests. The programme of action for the protection of tropical forests passed by the German parliament is an example worth mentioning, as it represents a most remarkable contribution on a national level. In this connection, a study group of both politicians and scientists has presented a comprehensive report "Schutz der tropischen Wälder. Eine internationale Schwerpunktaufgabe" (Bonn 1990, 983 pp.) (1).

Having been further highlightened by intensive scientific research and many campaigns by non-governmental organizations, the whole complex of tropical forests and deforestation meanwhile constitutes one of the major environmental problems of mankind. There is also, for the first time, a common consensus among scientists and politicians that tropical forests represent a unique and irreplaceable ecosystem, the destruction or even eradication of which would be a disaster for the global climate and environment. The extraordinarily high public awareness of the problem of tropical deforestation, which arose in the 1980s, is nowadays clearly to be seen all over the world, although effective steps and measures against tropical deforestation are still inadequate and weak.

The present paper attempts to review briefly the major aspects of the scientific discussion of tropical deforestation as far as the effects on climate and environment are concerned. All details presented in this paper were derived from the literature concerned, added to by the author's own experiences and observations in the field. An in-depth discussion, however, cannot be given, due to the limited space available. Although they are worth looking at in detail, a critical discussion is not possible of terms and definitions, such as: What are the tropics? What are forests? What is deforestation? These basic terms are used in the present paper in a rather loose sense and will need in future much more critical attention. Although tropical forests are commonly divided into closed and open forests, most of the comments usually found in the literature deal only with tropical forests as a whole. However, even the question whether only two types of tropical forests can adequately express the large number of existing tropical forest types and their different ecological valences remains open to dispute (2).

2. Extent of Tropical Deforestation in the Past and Future Projections

Among the alarming problems discussed, high priority is commonly paid to the rate and extent of tropical deforestation. Considering the whole earth, presently some 25 per cent, that is, around 35 mill. sq.kms, are under forest, out of which 16 to 17 mill. sq.kms are in the tropics, regardless of the different definitions of the tropics. This figure means that almost half of all tropical forests have already been cleared, irrespective of the reason (whether natural or man-made) and the purpose (whether for farmland by burning or for commercial timber exploitation or for fire-wood). Despite all inaccuracies in the methods applied for assessing the area under forest, it is commonly stated that the rate and extent of tropical deforestation have remained constant during the 1980s and even accelerated over time - despite great public interest in the problems concerned with tropical deforestation and a sincere public awareness of the great importance of the tropical forest ecosystem.

It is not possible to arrive at a reliable assessment of the tropical deforestation which has taken place in the recent past. However, the alarming extent of deforestation is nevertheless expressed by different estimates given in the literature. According to the FAO (1982), in 1980 tropical deforestation affected a total of 428,000 sq.kms, out of which primary rain forests accounted for 287,000 sq.kms (3). For the same year referred to (1980), LANLY (1982) estimated that a much larger total of 563,000 sq.kms of tropical forests had been cleared; this total consisted of 339,000 sq.kms of closed (primary) forests and 224,000 sq.kms of open (secondary) forests (4). Also referring to 1980, SEILER and CRUTZEN (1980) estimated the loss of tropical forests at between 305,000 and 794,000 sq.kms, of which about two thirds consists of closed forests and one third of open forests (5). The different estimates given by various authors and sources may result both from different definitions used for the tropics and forests as well as from different methods applied for assessing the areas under forest. Also, the term 'deforestation' may be defined in different ways. However, beyond any doubt, serious destruction and loss of forests is proved by all estimates.

Regarding the extent of tropical deforestation through the 1980s, MYERS (1989) has arrived at very alarming preliminary findings, estimating that for closed tropical forests the extent of deforestation increased by 90 per cent (6). In 1980, about 75,000 sq.kms of closed (virgin) tropical rain forests were cleared, and this figure increased in 1989 by 90 per cent to 142,000 sq.kms (7). This means a rate of 1.4 per cent for the annual deforestation of the existing lands under (close) tropical forests. On the basis of this estimate, the total forest cover in the tropics, in comparison to all tropical lands, dropped from about 40 per cent (for 1980) to a present-day estimate of only some 37 per cent, of which again about 40 per cent are open forests with a lower ecological valence than for closed tropical forests.

If the rapid process of tropical deforestation continues to accelerate, according to development in the 1980s the area under tropical forest will drop to only some 5 to 8 mill. sq.kms in the year 2050 (8). This means that only 30 to 50 per cent of the presently existing tropical forest will survive the next 60 years. However, it must be expected that the extent of deforestation will even be worse, due to the duplication of the world population and the availability of forests as almost the only resource then for the further extention of farmland. According to other estimates, the area under tropical forest will drop to 8 to 12 mill. sq.kms in the year 2050 (9).

The percentage figures for land under tropical forest, however, also vary between the continents and countries concerned. According to the FAO forest survey of 1980, the highest percentage of tropical forests was assessed for tropical America (with a figure of 53 per cent), with only 32 per cent for tropical Africa and 36 per cent for tropical Asia (10).

For the present-day situation, these estimates have further to be reduced by a tenth. According to the FAO forest survey, among the tropical countries also remarkably large differences occur in the percentage area under forest, showing the highest values for Guyana (95 per cent forest cover) and Surinam (93 per cent) while the lowest values are given for Niger, Burundi and Haiti with a forest cover of only 2 per cent of the total area in each country (11).

Estimates by MYERS, expressing the extent of tropical deforestation in 1989, show a very heavy loss of forests over a single year only, with the maximum values on the earth at as high as 15.6 per cent for Ivory Coast and 14.3 per cent for Nigeria (12). If this dramatic rate of deforestation continued, tropical forests in both countries would already be completely eradicated in the 1990s. As an alarming sign of the heavy present-day destruction of tropical forest, the extent of deforestation can be given for Brazil, which contains the largest forest cover among all tropical countries, showing a loss of some 50,000 sq.kms, or 2.3 per cent of all forests in Brazil, for the single year of 1989 (13). This means that, on average, every day nearly 150 sq.kms of forest were cleared. Other sources even estimate the annual extent of deforestation for Brazil at 80,000 sq.kms (14).However, the estimates of deforestation for Brazil vary considerably; for example, FEARNSIDE gives an estimate of about 25,000 sq.kms (15).

The rapid present-day destruction of tropical forests is a most alarming sign of a severe decline in future if no efficient counter-measures are taken. On the assumption that tropical deforestation continued to take place, forest land would drop to half of the present-day extent in 50 years and finally result in complete extinction by the end of the next century. However, it is urgently expected that, stimulated by a further rapidly growing public awareness about tropical forests, the protection of tropical forests will effectively take place through an international "action programme", together with efficient programmes of afforestation. On the other hand, afforestation should not be overestimated at the moment, because only a tenth of the total deforested area falls tentatively under current afforestation campaigns (16). Nevertheless, some countries have started agricultural practices which save forests from burning and clearing, such as eco-farming, which are hopeful signals pointing towards the preservation of tropical forests.

3. Reasons for Tropical Deforestation

Being a matter of serious concern and wide-ranging ecological harm, the drastic decline of tropical forests also represents one of the most serious ecological disasters on earth. If counter-measures towards the protection of tropical forests are to work successfully, it seems essential to reflect carefully on the manifold reasons for deforestation. In the light of the extensive references, including many case-studies from tropical countries, the following reasons need prior attention:

(1) The growing demand for farmland, under both permanent and temporary use, the latter representing the widely applied farm-system of shifting cultivation;

(2) the need for more pastures and grasslands for animal husbandry, both for intensive and extensive farming;

(3) the demand for land for settlement projects and industrial areas, including new roads and other infrastructural projects;

(4) the large extent of timber exploitation (partly illegally) from tropical forests for the purpose of building and construction materials in the home country and for exports, preferably to industrialized countries (by exploiting valuable timber like teak, ebony, mahogany, rosewood and sandalwood);

(5) the growing demand for firewood from forests, in order to satisfy the demand of the local people in the rural as well as urban areas for which fuel wood is the major, if not in fact the only, source of energy for domestic purposes and partly heating (in the 'outer' tropics).

All these reasons are, in fact, man-made, and most of them derive from the heavy burden of overpopulation and population pressure on land which result from the rapid growth of population. This basic reason seems to have been widely neglected so far and, thus, plays only a secondary role in the scientific discussion of tropical deforestation. On average for all tropical countries, the growth rate in population amounted to 2.5 per cent for the 1980s, which would result in a duplication of the population within less than 30 years. Because of the one-sided agrarian structure of most tropical countries, the massive increase in population subsequently has led to high pressure on the land available.

In order to satisfy the increasing demand for agricultural production, intensification and diversification of landuse on farmlands presently under cultivation would not be appropriate, so that the clearing of forests seems unavoidable for new and additional agricultural lands. On the other hand, most tropical countries are, due to their agrarian structure and their weak financial resources, not in a position to offer other than agrarian activities to the growing population. Deforestation, therefore, is in many cases the consequence of overpopulation and pressure on land. Consequently, population control and family planning must be seriously considered as effective instruments against deforestation.

4. Effects of Deforestation

Despite the controversial estimates of the extent and rate of tropical deforestation, there is a consensus that deforestation seriously affects the tropical forest ecosystem and, furthermore, also contributes towards a global change of climate and environment. Some common notions about the climatic and environmental impacts of tropical deforestation will be briefly reviewed (17).

4.1 Deforestation and Rainfall

There is widespread agreement that deforestation causes a reduction of rainfall on a local and even regional scale, because the recycling process of water vapour through forest transpiration is being reduced where there is deforestation. As a fundamental relationship, water vapour through forest transpiration increases with the biomass of forests, and subsequently also the processes of condensation and ultimately the formation of clouds and rain increase while it decreases with a lesser biomass.

Despite the essential validity of this relationship, it is disputed to what extent rainfall decreases through deforestation. Simulation studies of a large-scale transformation of the Amazon tropical rain forests into grassland have shown that the long-term annual total of rainfall, around 2,000 mm, would decrease by 11 per cent, or 220 mm, and furthermore, the 3-month dry season from June through August would be lengthened; due to both reasons, the evergreen tropical rain forests would no longer withstand the drier climatic conditions (18). It is generally accepted that these forests require a minimum of 1,600 to 1,800 mm rainfall per year and cannot resist a drought period of more than three months.

Other models of computing the potential decrease of rainfall, due to the transformation of tropical rain forests into farmland, have shown for the Amazon Basin that a transformation of 10, 20 and 40 per cent of the forests would reduce the annual rainfall total by 2, 4 and 6 per cent, respectively (19). Whether these findings are accurate or approximative, the common notion of a loss of rainfall through deforestation is nevertheless confirmed, giving, however, more serious evidence for local than regional climate.

4.2 Deforestation and Temperature

Although so far only given less attention only, the climatic impact of deforestation on temperature cannot be neglected; rather, a warming must be expected. The reason for this is the simple effect of transpiration cooling by plants, causing a drop in temperature; in the case of deforestation, however, transpiration cooling does not occur. So far, no definite results can be reviewed that show precisely the rate of temperature increase through deforestation. Like the impact on rainfall, the impact of deforestation on temperature also is a local rather than regional effect.

Nevertheless, in connection with tropical deforestation the greenhouse effect, intensified by deforestation, must also be considered.

4.3 Deforestation and CO_2

A prevailing consensus on the relationship between tropical deforestation and CO_2 evidently considers that deforestation through burning, which takes place on a serious scale with shifting cultivation, plays a major role in the increase of the atmospheric CO_2 concentration and, hence, in the greenhouse effect on an even global scale. Although the extent of the net carbon flux due to tropical deforestation is disputed, different estimates may be summarized by saying, that annually a net total of 1 +/- 0.6 billion tons of carbon (C) is emitted into the atmosphere by the burning of tropical forests. This massive volume of net carbon emission, which cannot be fixed in the forest biomass and therefore remains in the atmosphere, corresponds to between 7 and 32 per cent of all global CO_2 emissions released from the combustion of fossil fuels (20).

The above-mentioned figure is certainly rather vague and imprecise, but, nevertheless, it expresses the strong impact of tropical deforestation on the atmospheric CO_2 emissions. If the total net carbon flux into the atmosphere that results from tropical deforestation is considered for the continents, South America represents the largest source of CO_2 emissions from tropical deforestation, namely some 40 per cent, followed by Asia (37 per cent) and Africa (23 per cent). Among countries, Brazil and Indonesia lead with 20 per cent and 12 per cent respectively, thus producing together almost a third of all CO_2 emissions through tropical deforestation (21).

These briefly reviewed results show that the relationship between deforestation and CO_2 must be investigated in two respects: on the one hand, by studying the lesser biomass, due to deforestation, which has only a lesser ability to fix atmospheric carbon; on the other hand, attention must be paid to the increased volume of CO_2 emissions, due to tropical forest burning, mostly in connection with shifting cultivation.

4.4 Deforestation and Soil Erosion

Although erosion is a natural orogenetic process, which persistently changes the surface in all climatic zones of the earth, soil erosion is generally considered to be one of the most serious man-made impacts of deforestation on the environment. Forests, with their dense vegetation cover on the ground, naturally protect the soil from erosion in two ways: first, they act as an interception buffer, and secondly, they slow down the rate at which rain falls. After the clearing of a forest, however, the bare soil is exposed to erosion. The rate of erosion depends upon several factors, mainly the physical and chemical structure of the soil, the slope gradient and the intensity of rainfall.

As a result, principally caused by gravity, different types of erosion may occur, such as sheet erosion, splash erosion, landslips or even mass wasting (like landslides, creep, etc.). In consequence of all these types, the fertile top layer of the soil is washed away, which ultimately may result in bare land with poor environmental conditions for any substantial plant and animal life (22).

The rates of surface erosion for various tropical forest and tree crop ecosystems were given by WIERSUM (1984), who reviewed the results from 80 studies in the tropics (Table 1) (23).

Table 1 (figures are in tons per hectare and year):

Type of ecosystem	Minimum	Median	Maximum
Natural forests	0.03	0.3	6.2
Shifting cultivation, fallow period	0.05	0.15	7.4
Shifting cultivation, cropping period	0.4	2.8	70.0
Forest plantation, undisturbed	0.02	0.6	6.2
Multi-storeyed tree gardens	0.01	0.06	0.14
Tree crops with cover crop mulch	0.10	0.75	5.6
Agricultural intercropping in young forest plantations	0.6	5.2	17.4
Tree crops, clean weeded	1.2	47.6	182.9
Forest plantation, litter removed/burned	5.9	53.4	104.8

Although these values have to be understood with some restrictions (for example, because of the neglected pedological situation), it nevertheless becomes evident that surface erosion is minor in ecosystems where the top soil is adequately protected against the impact of rain drops by a well-developed close vegetation cover, particularly by a dense understorey. On the other hand, the removal of the understorey or even destruction of the litter layer dramatically increases the rate of erosion. Needless to say, litter gathering, as a substitute for firewood, and forest burning intensify the risk of soil erosion while certain agricultural practices like mulching are of a beneficial effect, because they prevent erosion. It is also noteworthy to see that a tree community, without an understorey, is very adversely affected by soil erosion.

4.5 Deforestation and Floods

Mainly caused by clearings of forests in tropical uplands, severe floods have been very frequently recorded and reviewed in many studies. Due to the low infiltration capacity of the soil, in combination with high rainfall intensities (which are characteristic of tropical rainfall), extremely high rates of run-off occur which may significantly increase the stormflow volumes of rivers and streams. Even for smaller rivers, significant increases in stormflow have been observed, depending upon the extent of deforestation of the catchment area.

Floods occur more seriously in the case of complete forest clearings, but many examples of severe flooding have also been given for tropical upland areas which were transformed from forests into plantations, such as tea.

Besides the low or even non-existent infiltration capacity of the soil, the non-existent rate of transpiration in the case of forest clearings also largely contributes to the risk of flooding and stormflow. It is commonly noted that in tropical forests up to 50 per cent of all rainwater evaporates from the leaves of trees and plants.

The majority of floods caused by deforestation are also partly due to the large quantities of silt sedimentation in the tropical lowland streams, through which the streambed and watertable are lifted and hence the risk of floods is increased. Therefore, the higher rate of soil erosion through upland deforestation is closely linked with the greater number of downstream floods. Numerous investigations clearly prove the large sedimentation loads in downstream rivers.

The ecological impacts of deforestation so far reviewed must be added to by other adverse effects which, however, cannot be further discussed, such as the irreversible destruction of the unique diversity of plant and animal species and thus the depletion of the earth's genetic resources. The effects of deforestation by burning on trace gas emissions, which will not receive further comment also need much more attention.

5. Suggestions

Since the tropical forest ecosystem is by its very nature highly vulnerable and since it is, in practice, adversely affected in many respects and regions, all efforts have to be made for an immediate international "action programme" towards the conservation and preservation of tropical forests, but also towards afforestation. It seems doubtful, however, whether such an "action programme" can be successfully conducted without being actually imposed with rational landuse practices, such as eco-farming and agroforestry, particularly in densely populated regions, and also compelled with strict family planning policies and programmes, in order to reduce the serious pressure on land.

Although tropical forests are obviously located in tropical, or rather Third World countries, these countries cannot carry out the heavy burden of forest protection and afforestation alone, but need the support and collaboration of developed countries and international organizations. The protection of the tropical forest ecosystem is an international task, but also a challenge towards international solidarity between the rich and poor countries. Therefore, all efforts must be considered as a contribution towards the protection of the global environment and thus the survival of mankind.

6. References

/1/. The report, containing a list of about 1,500 references, represents a most substantial compilation of publications on tropical forests and deforestation, and is part of the study group's work on "Preventive Measures to Protect the Earth's Atmosphere".
/2/. WALTER,H.: Vegetation and Klimazonen Stuttgart 1979.
/3/. FAO: Tropical Forest Resources. In: FAO Forestry Research Paper No.30, Rome 1982.
/4/. LANLY, J.P.: Tropical Forest Resources. In: FAO Forestry Research Paper No. 30, Rome 1982.
/5/. SEILER, W. and P.F. CRUTZEN: In: Climate Change Vol.2, 1980.
/6/. MYERS,N.: Deforestation Rates in Tropical Forests and their Climatic Implications. A friends of the Earth Report, London 1989.
/7/. FAO: An Interim Report on the State of Forest Resources in the Developing Countries. Rome 1988.
/8/. MYERS, 1989: Schutz der tropischen Wälder. Eine internationale Schwerpunktaufgabe. Zweiter Bericht der Enquete-Kommission des 11. Deutschen Bundestages "Vorsorge zum Schutz der Erdatmosphäre", Bonn 1990.
/9/. BRÜNIG, E.B.: Die Erhaltung, nachhaltige Vielfachnutzung und langfristige Entwicklung der tropischen immergrünen Feuchtwälder (Regenwälder). Arbeitsbericht, Bundesforschungsanstalt für Forst - und Holzwirtschaft, Institut für Weltforstwirtschaft und Ökologie, Hamburg, 1989.
FAO (1990) in /8/.
/10/. FAO (1888).
/11/. FAO (1988).
/12/. MYERS (1989).
/13/. MYERS (1989).
/14/. GTZ Info. No. 4, 1989: Verändert sich das Weltklima? pp. 12-18.
/15/. FEARNSIDE, P.M.: in print as cited in /8/.
/16/. FAO (1988).
/17/. Among others see PREM SAMPURNO BRUIJNZEEL: Environmental Impacts of (De)Forestation in the Humid Tropics. A Watershed Perspective. In: Wallaceana, Vol. 46, pp. 3-13, 1986.
/18/. HENDERSON-SELLERS, A. and V. GORNITZ: Possible Climatic Impacts of Land Cover Transformation, with Particular Emphasis on Tropical Deforestation. In: Climatic Change Vol. 6, pp. 231-257, 1984.

/19/. BROOKS, K.N.: Evaluation of Deforestation Impacts on Environment and Productivity. In.: Proceed. IXth World Forestry Congress, Mexico, E-1.6.1.A, 13 pp., 1985.

/20/. MELILLO, J.M. et al.: Land-use Change in Soviet Union between 1850 and 1980: Causes of a New Release of Co_2 to the Atmosphere. In: Tellus Vol. 40 B, pp. 116-128, 1988.

/21/. Schutz der Erdatmosphäre. Eine internationale Herausforderung. In: Berichte der Enquete-Kommission des Deutschen Bundestages "Vorsorge zum Schutz der Erdatmosphäre", Bd.1, Bonn 1990 (3rd ed.), p. 540.

/22/. BRÜNIG, E.F.: Die Entwaldung der Tropen und ihre Auswirkungen auf das Klima. 1987.

/23/. WIERSUM, K.F.: Surface Erosion Under Various Tropical Agroforestry Systems. In: C.L. O'Loughlin and A.J. Pearce (Eds.): Proceed. Symp. on Effects of Forest Land Use on Erosion and Slope Stability, 1984. Honolulu, pp. 231-239, 1984.

Energy and Atmospheric Environment in Post-Communist Countries: The Case of Poland

J. Paszynski

Institute of Geography and Spatial Organization,
Polish Academy of Sciences, ul. Krakowskie Przedmiescie 30,
PL-00-927 Warsaw, Poland

Deep transformations of political, social and above all economic systems in the post-Communist countries of Central and Eastern Europe, involve consequences for the structure of demand, production and supply of energy. It is extremely difficult to analyze these changes for two reasons:

- The changes mentioned are still in progress.
- These changes are different in each country.

The main focus of this contribution is on Poland, the country where political change first began ten years ago when "Solidarity", the first independent trade union in the Communist block, emerged. The recent transformations are probably most profound in Poland, with regard to the national economy at least, and their effects are of great importance for the structure and efficiency of energy systems. Remarkable changes in the Polish economy have taken place in the last year since the abolition of the totalitarian, communist regime. The main consequence of this political change is the replacement of the centrally planned and centrally ruled economy by a market economy.

The population of Poland is about 38 million and it is growing rapidly in contrast to some other post-Communist countries, such as Czechoslovakia or Hungary. This growth during the last decade has been approximately 0.8% yearly. In this connection, the total consumption of primary energy in Poland increased from 173 million tons of coal equivalent (TCE) to 181 million TCE in the period 1978-1988, but its consumption per capita decreased by about 3.5%.

The main sources of energy in Poland are fossil fuels. The share of fuels in the primary energy input in 1988 was as follows:

hard coal 68%
lignite 11%
crude oil 14%
natural gas 7%

This structure changed considerably during the period 1978-1988. The share of lignite almost doubled from 41 million tons in 1978 to 73 million tons in 1988, while the share of crude oil fell to some extent from 20 million tons in 1978 to 17.5 million tons in 1988. This structure of primary energy inputs involves a relatively low yield of final energy.

Coal and lignite are of domestic origin while almost the entire amount of crude oil and its derivatives are imported. The annual output of hard coal is about 190 million tons, but almost one fifth of this amount is exported, being an

The Global Environment

Editors: K. Takeuchi · M. Yoshino © Springer-Verlag Berlin, Heidelberg 1991

important source of badly needed hard currency. The quality of coal provided for export is much better than the quality of coal used for domestic purposes. The average calorific value of coal exported is 26 GJ/ton, compared with 18-22 GJ/ton for coal used domestically.

Of the total energy consumption, the share of industry is about 41%, that of public transport, railways and public road transport about 4%, and that of fuel mining about 4%. The remaining 51% is consumed by residential and tertiary sectors of the economy. Energy consumption in Poland is very inefficient. Its use for the same output is on average 30-60% higher than in industrialized West European countries.

Combustion of fossil fuels for the generation of energy releases air pollution in the form of solid, liquid and gaseous products. One of the pollutants produced by energy generating systems is CO_2. Total emissions of CO_2 from all fossil fuels has been estimated as high as 120 million tons in 1988. About 72% of this amount is due to the burning of hard coal, 12% to crude oil, 11% to lignite, and the remaining 5% to natural gas. These estimates are based on the following emission coefficients: 23.9 kg CO_2/GJ for solid fuels, 19.7 kg CO_2/GJ for oil, and 13.8 kg CO_2/GJ for natural gas. Industry and fuel mining are responsible for 46% of the CO_2 emissions while the rest is attributable to transport and domestic heating.

More detailed data on atmospheric pollution in Poland concern sulphur components. The sulphur content of the best quality bituminous coal, which is exported, amounts to 0.7 - 1.0%. Coal provided for internal use is of much poorer calorific value and much higher sulphur content, up to 2%. The sulphur content of Polish lignite is 0.8 - 1.0%. The total amount of SO_2 emissions from all sources in Poland is now reaching about 4.3 million tons/year. The corresponding value for East Germany, the former GDR, is still higher: 5.2 million tons/year. The main source of SO_2 emissions in Poland, accounting for 48%, is the combustion of solid fuels, both bituminous and lignite, in combined heat and power stations (CHPS). Polish industry is responsible for 26% of total SO_2 emissions, and combustion in small individual domestic stoves for another 23%. The rest, 3%, is emitted by motor vehicles.

Poland is one of four European countries where the emission of SO_2 is still growing. During the decade 1978-1988, these emissions increased by about 10%. A very important role in atmospheric pollution in Poland is played by the transboundary flow of SO_2. The inflow of SO_2 from adjacent countries, mainly from Czechoslovakia and East Germany, is a very serious problem. Poland obtains more than 0.6 million tons/year from the former GDR and almost 0.3 million tons from Czechoslovakia. The inflow of SO_2 to Poland is estimated as being a little higher than the outflow of SO_2 from Poland. The mean annual value of this inflow is 1.2 million tons. In consequence the concentration of SO_2 in the atmosphere as well as its deposition attain their highest values in the south-western part of Poland, in regions adjacent to Czechoslovakia and Germany, chiefly in Silesia.

A similar situation occurs in Czechoslovakia, where total annual emissions of SO_2 are as high as 3 million tons. The deposition of sulphur from domestic sources is only 52% of the total amount of deposited sulphur. The transboundary inflow from East Germany is responsible for another 20%, from Poland for 10%, and from Hungary and West Germany for 6% and 5% respectively, according to recent estimations made by Slovak meteorologists.

The south-western part of Poland is not only exposed to very important transboundary transport of atmospheric pollution but also is the region where the highest emissions of SO_2 occur. The situation is especially dangerous in the Upper Silesian Industrial District, where a large part of Polish industry is located. As a consequence, the highest concentrations of atmospheric SO_2 are observed in this region.

Mean annual concentrations of SO_2 in Poland's rural areas are between 10 and 15 $\mu g m^{-3}$. The south-western part of Poland has average concentrations over 30 $\mu g m^{-3}$, and the big industrial agglomerations of Upper Silesia and of the Cracow region have mean SO_2 concentrations as high as 150 $u m^{-3}$. Other places of very high SO_2 concentration are areas adjacent to big lignite mines, chiefly those where large lignite power stations have been built, e.g. Turow or Belchatow in the central part of Poland in the vicinity the Lodz conurbation.

Analysis of mean annual values of SO_2 concentrations in Poland for the decade 1980-1990 reveals that minimal concentrations occurred in 1981, the year when martial law was imposed. Another reduction of SO_2 concentrations was observed during the last two years. It is very probable that both cases are the consequence of decreased emissions of SO_2 due to reduced energy consumption brought about by political crisis and economic depression during those periods. It is certain that industrial production in 1989-1990 diminished by 20-30%, resulting in reduced energy consumption compared to the previous period.

One of the consequences of SO_2 emissions is increased acidity of atmospheric precipitation, measured as pH. Mean values of pH of precipitation in 1987 were 4.5 in north-east Poland and 4.3 in central Poland. The lowest pH values for rain water have been observed in south-west Poland, in the Sudety Mountains, where their average value in 1987 was as low as 3.8, according to data from Sniezka Peak.

In conclusion Poland is one of those countries which have very high emissions of the gaseous atmospheric pollutants, CO_2, SO_2. This is due to the unfavorable structure of Polish industry, inherited from the communist centrally controlled system, which placed great emphasis on the extremely energy-consuming sectors such as mining, the steel industry, and heavy engineering. The same industrial structure is typical of other post-communist central European countries, especially Czechoslovakia and East Germany. Another factor playing an important role is the lack of energy-saving mechanisms, again the consequence of the state-owned and state-monopolized economic system in these countries.

Very bold, radical, and also very costly steps, in conjunction with substantial transformation of the entire political, social and economic system, even requiring changes in people's attitudes, are necessary to solve our environmental problems. A high level of international cooperation would seem to be essential, if effective change is to occur.

Further Reading

Analytical Review of Background Environmental Pollution in the Region of Eastern European CMEA Member-Countries, 1980-1986. Gidrometeoizdat, Moscow 1988, 73p.

Ochrona srodowiska Preservation of Environment. Glowny Urzad Satystyczny - Materialy statystyczne, 68, Warszawa 1989, 223p. (in Polish)

Sitnicki, S., A. Szpilewicz, J. Michna, J. Juda, K. Budzinski: Greenhouse Gas Control Strategies - Case Studies in International Cooperation - Poland: Opportunities for Carbon Emission Control. Warszawa 1989, 41p. (typescript)

Zavodsky, D., K. Pukancikova: Long range transport and deposition of sulphur and nitrogen compounds on the territory of Czechoslovakia in the years 1985-1987. Meterologicke Zpravy, 42, 1989, 5p. 147-151. (in Slovak)

Nowicki, M. Stan zanieczyszczenia atmosfery w Polsce w 1987 roku Atmospheric Pollution in Poland in 1987. Instytut Ochrony Srodowiska. Warszawa 1990. (in Polish)

Influence of a Global Warming on the Risk of Forest Fires in the French Mediterranean Area

A. Douguédroit

Institute of Geography, University of Aix-Marseilles II,
29 avenue Robert Schuman, F-13621 Aix en Provence Cedex, France
and Environment Climatique (C.N.R.S.),
B.P. 53, F-38041 Grenoble Cedex, France

Abstract. The important forest fires which spread in the French Mediterranean area can be explained by a high combustibility of the vegetation due to little soil moisture. To estimate summer drying, the potential and real evapotranspiration (P.E.T and R.E.T.) have been calculated for Marseilles area with several hypothesis of temperature and precipitations to take into account natural variations and the direct influence of a global warming. Some predictions of variations of soil moisture by General Circulation Models when forced by an increase of carbon dioxide have been investigated.

Résumé. Les grands feux de forêts qui éclatent dans la région méditerranéenne française supposent une grande combustibilité de la végétation causée par une faible humidité du sol. Pour estimer la sécheresse estivale, on a calculé l'évapotranspiration potentielle et réelle dans la région de Marseille pour plusieurs hypothèses de température et précipitations, tenant compte des variations naturelles et de l'influence directe d'un réchauffement climatique. Quelques prédictions des variations de l'humidité du sol faites par des modèles de circulation générale forcés par un accroissement du CO_2 ont été étudiés.

Key-Words. Forest fires - Evapotranspiration - Mediterranean climate - Influence of Global Warming.

Mots-clés. Feux de forêt - Evapotranspiration - Climat méditerranéen - Influence du réchauffement climatique.

Forest fires are considered as a major disaster in the French Mediterranean area. Their importance, which has been well known for several decades, has been measured more accurately since 1973, when a particular office was created to follow the evolution of the problem. From 1973 to 1989, in the administrative districts of the Mediterranean coast (44704 km² : Fig. 1), 3177 km² have burnt, mostly during summer (69 % out of the total in July and August), not to

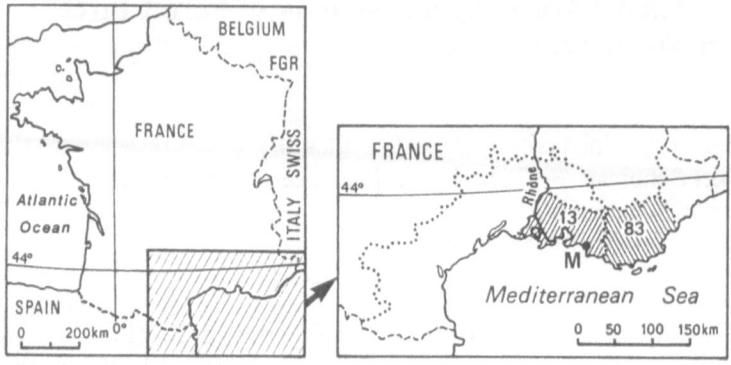

Fig. 1 - France and the French Mediterranean area -Northern limit of administrative districts
13 : Bouches-du-Rhône   83 : Var   M. : Marseilles

mention fires smaller than a hectare. Huge forest fires are famous, the most important of them spread over 18 000 and 12 000 hectares or over famous sceneries as Mount Sainte Victoire. In total, 7.1 % of the whole surface has been affected by fires for seventeen years. A likely effect of the predicted global warming due to an increase of carbon dioxide (CO_2) in the atmosphere in the next decades makes it urgent to investigate its influence on the climatic condition prevailing in the breaking out and the spreading of forest fires. The results should be taken into account in a possible preventive policy.

1. Factors favoring forest fires

Forest fires only spread when vegetation is dry ; the high combustibility of leaves, branches and trunks connected with water deficiency in soil moisture, is the result of climatic evolution during previous months. But their spreading, after breaking out, depends on local conditions such as air humidity and mainly wind. Strong winds like the Mistral and the Tramontane, which frequently occur in the French Mediterranean area are the most important factors favoring the spreading of fires when vegetation is dry [1-3]. They are common in summer in the Mediterranean climate at the same time as low air humidity and little soil moisture. Climatic indices used for predicting the spreading of fires are not investigated here.

1.1. Soil moisture

Soil moisture is very difficult to determine with a reliable accuracy. Measures are either directly unusable when done with blocks of porous porcelain or recent and rare when done with neutron probes because of difficulties in their management. Most of the time, soil moisture is estimated by formulas which are well known nowadays. They are based on two notions expressing water loss to the atmosphere. The loss-rates are known as potential evapotranspiration (P.E.T.) and real evapotranspiration (R.E.T.) rates.

P.E.T. can be considered from two points of view, for vegetation or climate. For plants, P.E.T. represents the evapotranspiration tending towards its maximum, owing to sufficient soil moisture drawing water to the roots without limit. Its variations depend both on energy and dynamic characteristics of regional environment on the one hand and on the period of the vegetative cycle of the vegetation on the other hand. From the climatic point of view, P.E.T. represents the energy required by the atmosphere, that is to say the present evaporation power of atmospheric environment expressed in terms of energy. In fact, it varies between two values, the maximum value of the real E.T. of a large vegetal shelter being a minimum, and the intrinsic evaporing air power without evaporation being a maximum because the energy required by the atmosphere can surpass the maximal water vapour flow of plants.

R.E.T. represents the variations of water vapour flow from plants. If it is higher than precipitations, it can be considered as equal to P.E.T. But if the energy required by the atmosphere increases or if the water available in the soil decreases, the exchanges of water between plants and the atmosphere is reduced and R.E.T. becomes lower than P.E.T.

1.2. Estimation of P.E.T. and R.E.T.

The different formulas to assess P.E.T. can be divided into two types : mathematical adjustments for THORNTHWAITE [4-5] and TURC, energy balance for PENMAN [6-7], whose method has been simplified for France by BROCHET and GERBIER from METEO FRANCE [8-11]. A comparison between measured values of P.E.T. in an experimental field and calculated ones done in the south of France during 3 years has proved that the empirical formulas are badly adapted to the Mediterranean climate. They notably underestimate P.E.T. in summer. PENMAN'S formula gives more satisfactory results and simplified formulas of METEO FRANCE are very close to the results of PENMAN'S [12].

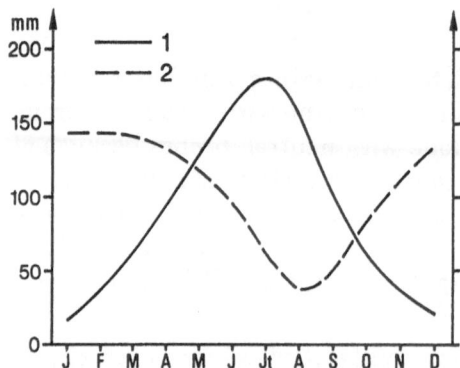

Fig. 2 - Global variation of potential evapotranspiration (above) and of soil moisture (below) in the French Mediterranean area.

So the most convenient formulas of P.E.T. to be used for the French Mediterranean area are based on the energy balance which represents a climatic interpretation of P.E.T. As the energy balance commands the temperature, variations of P.E.T. are connected with variations of temperature.

Real evapotranspiration is reduced during several summer months in Mediterranean climates when the hottest temperatures and low or even missing precipitations coincide. Soil moisture retention decreases down to zero or almost zero (Fig. 2). So the variation of R.E.T. depends on the energy balance when soil moisture is sufficient ie enough precipitations and on the other hand depends mainly on precipitations when they become rare.

Soil moisture which is the basic factor of forest fires is influenced by two main climatic elements, the energy balance which commands temperature and precipitations which regulate soil moisture. They prevail in turn, according to the season, the first one in winter and the second one during summer.

1.3 . Air humidity and wind

Air humidity depends on the nature and evolution of the air mass concerned. During summer, it is low when the air persists over a hot and dry continent. It makes a quick spreading of forest fires easier. Its variations are linked with the variations of synoptic situations and of temperatures.

As for strong winds, they are due to the combined influence of synoptic situations and local relief canalizing air flow in valleys (TABLE 1) . The possible influence of global warming is indirect.

TABLE 1

Mean annual number of occurrences with maximum speed of the Mistral reaching a given speed (1957-1981).

	J	F	M	A	M	Jn	Jl	A	S	O	N	D
≥ 10m/s	10.5	10.5	12.2	12.2	9.8	10.2	11.5	11.0	8.4	8.4	10.2	10.6
≥ 15 m/s	7.6	6.7	7.7	8.8	6.2	6.5	7.8	6.8	4.5	5.8	6.2	7.2
≥ 20 m/s	4.0	3.6	3.8	4.8	3.1	2.6	3.3	2.5	1.9	2.7	3.4	3.9
≥ 25 m/s	1.7	2.0	1.6	2.1	1.0	0.6	0.8	0.8	0.5	0.7	1.5	1.3

2. Influence of variations of temperatures and precipitations

Respective influence of variations of temperature and precipitations will be investigated through their influence on P.E.T. and R.E.T. We focused our attention on the area of Marseilles, which can be considered as representing general conditions of the two districts of Bouches-du-Rhône and Western Var including mainly areas of calcareous rocks covered with thin soils (Fig. 1 : 13 and 83). Climatic years beginning in October when rain becomes important and ending after the dry season in September will be used.

2.1 . Variations of P.E.T.

METEO FRANCE'S second formula has been used to calculate P.E.T. every ten days. It needs insolation, mean minimal temperature, mean temperature and mean wind speed every ten days. Data of the station of Marignane, the airport of Marseilles, were used (1951-1980).

The annual amount of P.E.T. calculated with mean thirty year values reaches 554 mm. Its monthly variations are characterized by an important difference between winter and summer (Fig. 3). Values during the hot season combine the effect of the increasing radiation and of the growing season of vegetation, and there is no telling them apart. When mean temperatures increase, P.E.T. rises too but only a little (about 10 mm for every additional degree). In the thirty year series 1951-1980, differences between mean and maximal annual temperatures of the stations of the area are one degree and between mean and maximal temperatures of summer months like July 2 or 3 degrees.

The predicted global warning due to the doubling of carbon dioxide is on average $3° K \pm 1.5$ according to the general circulation model (GCM)

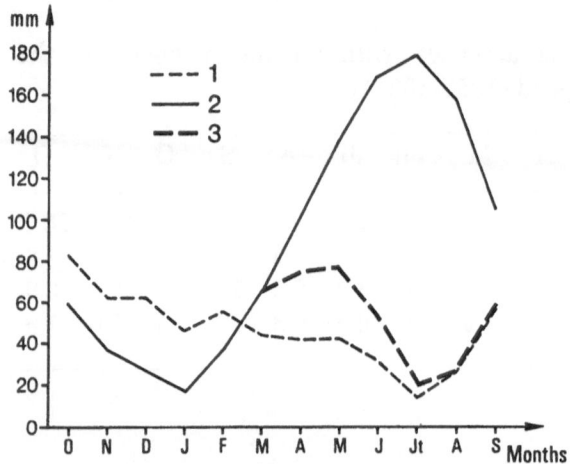

Fig. 3 - Monthly distribution of potential and real evapotranspiration in the area of Marseilles.
1 : Precipitations (mean 1951-1980)   2 : P.E.T.   3 : R.E.T.

chosen, and predictions are supposed not to be significant, from a statistical point of view, for the Mediterranean area.

If we only consider global warming itself, it could induce an increase of potential evapotranspiration as important as natural variations of temperature or a little more important but with a similar monthly distribution of P.E.T. during the year.

2.2. Variations of R.E.T.

R.E.T. is deduced from P.E.T. and precipitations. To get a longer series (from 1864), precipitations of the oldest station of Marseilles, the Observatory, has been used. It has not a normal distribution but a fairly dissymetric one (TABLE 2). Temperatures of stations of central Marseilles were not accepted because they are influenced by the urban heat island effect of a big city (a million inhabitants).

The decreasing of R.E.T. for months with precipitations lower than P.E.T. was calculated owing to usual soil moisture retention tables.

Several cases were investigated : mean values of precipitations of 1951-1980 and values of particular years (1988-1989 which is one of the driest in the long series of Marseilles, 1935-1936 the wettest with 1931-1932 which presents the particularity of a rather wet summer).

Every year can be divided into two parts : one from October to February or March, and the second till September. During autumn and beginning of winter, precipitations are equal or more important than P.E.T.

TABLE 2

Characteristic values in mm of the long series (1864-1989) of annual precipitations of Marseilles-Observatory (from October to September).
1 = Mean ; 2 = Standard-deviation ; 3 = Minus two standard-deviation value ; 4 = Absolute minimum ; 5 = Median ; 6 = Upper value of the lower tenth of the series ; 7, 8, 9 = Annual precipitations of 1988-89, 1935-36 and 1931,32.

1	2	3	4	5	6	7	8	9
585	158	269	283	570	390	311	1010	1088

The available part of the soil moisture which was non-existent at the end of summer is generally supposed to be reconstituted at the end of winter. Then R.E.T. equal P.E.T. (Fig. 3). During the end of spring and summer, R.E.T. decreases and becomes very lower than P.E.T.

Using the two groups of values of temperatures (mean 1951-1980 and mean +1°C) and corresponding P.E.T. mentionned before, rather low differences between corresponding values of R.E.T. were obtained. Higher temperatures infer lower R.E.T. but the difference is not very important and monthly distribution of R.E.T. is not modified. The difference increases when R.E.T. is calculated, the same values of temperature being used, for the driest and the wettest years. With 311 mm in 1988-1989 (the fourth driest year) R.E.T. decreases in more than 140 mm by comparison with a year with mean precipitations. With 1010 mm of rain in 1935-1936, it does not increase a lot (Fig. 4).

The most important differences between the annual calculated amounts of R.E.T. depends on the repartition of precipitations over the twelve months. A very rainy year means very rainy autumn and winter. But when the water holding capacity is full, rain supplies runoff and R.E.T. remains very much the same. When summer is wetter, R.E.T. increases very much. In 125 years in Marseilles, rain exceeded 50 mm in July eight times and 100 mm twice. With 271 mm during hot season including 141 in July, and 274 in September, 1932 was one of the wettest years of the series and 1931-1932 the wettest. It induced a growth of R.E.T. that exceeded 230 mm, particularly in summer when R.E.T. is closely linked to the distribution of precipitations (Fig. 5).

The influence of variations of precipitations on the available water in the soil depends on their distribution over the year. A wet rainy season does not much modify the importance of R.E.T., except if it is followed by an exceptional wet summer. But a dry rainy season induces at its

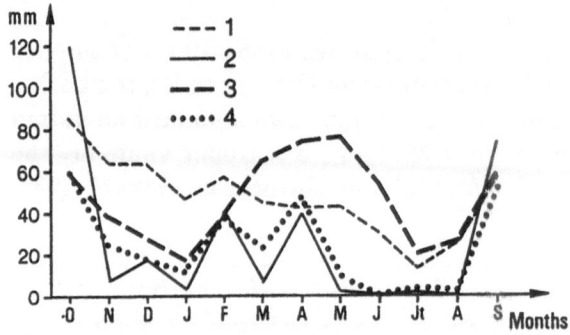

Fig. 4 - Monthly distribution of R.E.T. in the area of Marseilles. Cases of 1951-1980 and 1988-1989
1,2 : Precipitations 1951-1980 and 1988-1989
3,4 : R.E.T. 1951-1980 and 1988-1989

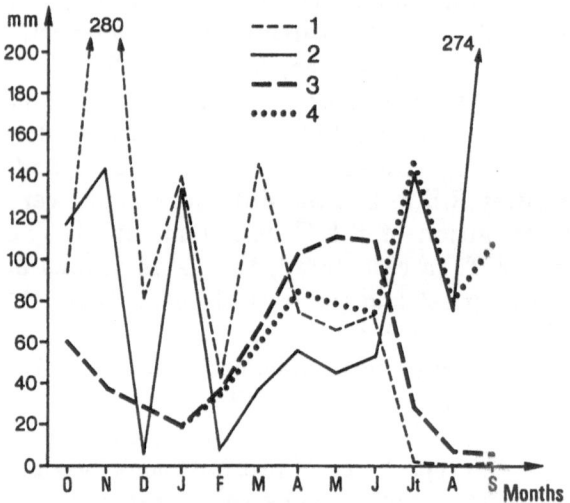

Fig. 5 - Monthly distribution of R.E.T. in the area of Marseilles. Cases of 1935-1936 and 1931-1932
1, 2 : Precipitations
3,4 : R.E.T.

end soil moisture lower than field capacity ; it reduces R.E.T. from the end of spring and makes the dry season longer for vegetation threatened by forest fires. It induces a more important modification in soil moisture than a weak increase of temperature.

3. Comparison with GCMs' predictions

All attention is nowadays focused on the direct effects of global warming and neglects possible indirect effects. These effects are taken into account in the GCMs which simulate variations of soil moisture including unavailable water for plant roots. But the evolution of the socio-economic system could slightly modify all those rough approaches.

3.1. Possible indirect effects of global warming

Summer dryness is linked with the Azores High and high geopotentials in altitude up to and higher than 500 hPa. North Atlantic Oscillation (N.A.O.) synthetizes their variations over the Atlantic Ocean North of equator [13-15]. Its anomalies are connected (many climatologists would say are controlled) with anomalies of the surface temperature of the ocean (S.S.T.), the origin of which being unknown. Variations of N.A.O. greatly affects, in Western Europe and the Mediterranean Basin, variations of precipitations, through positions of the tracks of depressions, [16].and variations of temperature through the types of the air mass [17].

On the other hand, the existence of the Mistral is controlled by the repartition of pressure in the same area which is connected with the distribution of pressure over the Atlantic Ocean.

Unequal warming of the atmosphere over the planet must be associated with anomalies of intensity and of position indeed of areas of maximal and minimal temperature in the atmosphere. The chain reaction due to atmosphero-occan inter-relations supposes induced anomalies in S.S.T. Such anomalies influence soil moisture-cloud cover-precipitation feedback included in the GCMs and affect climatic characteristics over Western Europe and the Mediterranean Basin.

3.2. Predictions of the GCMs

GCMs predict the repartition of soil moisture for climates simulated with present (control) and increased atmospheric carbon dioxide. But it is a secondary feature of the models and one of the most difficult parameters to calculate because depending on a complex sequence of interactions associated with the hydrologic cycle. It is much influenced by all the modifications in parametrizing any component of the cycle, especially the cloud cover which is very simply characterized in most of the climatic models. Soil moisture is generally computed by the so-called "bucket" method. So, with an increase of carbon dioxide, regional effects on soil moisture predicted by the GCMs present very important differences.

The results of experiments based on a doubling of CO_2 in five climate models in which the regional changes in soil moisture distribution over North America was calculated, have been compared. For large-scale regional changes, there was little agreement about winter and a poor one about summer. Only partial explanations of the differences have been found [18-19].

If we look to GCMs' predictions concerning the Northern Mediterranean Basin, similar differences can be found. A comparison between the NCAR (National Center for Atmospheric Research) and the GFDL (Geophysical Fluid Dynamics Laboratory) GCMs' predictions present a good example of the uncertainty of results obtained [20-26].

With an increase of CO_2, both models predict no noticeable variation of water in the soil during winter and a decrease in summer. The decrease is less than 10 mm in the GFDL model ; it is supposed to be significant at 10% level when 4 x CO_2 or 2 x CO_2 according to different parametrization of cloud cover. It is significantly less in the NCAR model with a similar cycle. The difference between the results of both models lies in the amount of soil moisture in late winter and spring in the control case. This amount is near saturation in the GFDL model and lower in the NCAR one. In the doubled CO_2 case, it increases less in the first model during late winter and spring than in the second one; a greater evaporation rate results in a soil moisture deficit almost as soon as the warm season begins. In the NCAR model, drying is delayed.

The determining factor of the sensitivity of soil moisture to a warming due to an increase of CO_2 is the amount of water in the soil during late winter and spring. Differences between results obtained by the two types of models are similar to differences induced by natural variations of precipitations (see 2.2). The NCAR model has the same effect as an excess of rain during a cold season without any influence on the length of the dry season. The GFDL model induces a dryer summer season as do poor precipitations, but dryness is not as marked as when they are the poorest.

Right now, we must recognize inadequacies of the models to calculate soil moisture, and to determine how precipitations and soil moisture will change with an increasing atmosphere carbon dioxyde.

3.3. Influence of the socio-economic environment

The importance of huge forest fires is a rather new phenomenon. It was almost unknown before the twentieth century. It is related to the evolution of the socio-economic system in France which led to large areas deserted during the second half of the nineteenth century and the beginning of the present century. Before, forests and scrublands were used by farmers, particularly as pasture for sheep and goats. And then, their whole surface was quite smaller, because many cultivated terraces covered gentle slopes. They were given up when agriculture limited to low plains to develop intensive farming and are nowadays overrun by pines.

Such an evolution has consequences favoring forest fires. Low grass not eaten by flocks stand up and become dry fuel for fires. Pine cones contribute to fire spreading when they burst. Lanes are overgrown with vegetation and moving becomes difficult in the countryside ; that is to say access to fires is more difficult. New paths have to be opened. Farmers do not travel through forests and scrublands which are empty of people who could fight those fires which has just broken out. Specialized fire brigades arrive from far away when fires have developped due to strong winds.

Two facts confirm the importance of the socio-economic situation of the country. There are no important fires in North Africa because of the density of the population in countryside : forests are pastured and fires are quickly stopped by local people. During the nineteenth century in France, there were years with precipitations as rare as 1988-1989 (1876-1877, 1877-1878 and 1881-1882 for example) but without fires as large as nowadays.

The origin of forest fires is a very controversial question. It is often said, without any proof, that hunters disagreements in inner Provence and tourist resort planning along the coast are at the root of many fires. There is no doubt that regulations forbidding building for decades in areas where the forest has been burnt down would put a stop to public rumour.

The economic situation in the countryside in the South of France has created conditions in favor of the spreading of fires. Coming back to previous farming is out of the question. But a better policy of fire prevention and fire fighting can be organized to minimize fire risks. The combined possible effects of a global warming and of natural anomalies should increase the interest of such a new policy.

Conclusion

The important forest fires which spread out on French Mediterranean area during summer suppose a high combustibility of vegetation associated with a little soil moisture. The climatic year is divided into two main seasons : the rainy season with R.E.T. (Real Evapotranspiration) equal to P.E.T. (Potential Evapotranspiration) and the dry one when it is lower.

The effects of a global warming due to an increase of atmospheric carbon dioxide is difficult to assess on soil moisture. Variations of P.E.T. and R.E.T. directly induced by the direct increase of temperature can be calculated. But its indirect effects are rather hard to estimate. Predictions of variations by GCMs when forced by modifications in the heat balance of the atmosphere change because soil-moisture is a parameter very hard to calculate. But the possible combined effects of a global warming and of natural anomalies of precipitations should increase the interest of a new policy of fire prevention and fire fighting.

References

1. J.C.Barescut : La Météorologie, 26, 208 (1981)
2. P. Carrega : Rev. Analyse Spatiale, 24, 165 (1988)
3. J.C.Drouet, B. Sol: R.G.S., 90, 60 (1990)
4. C.W. Thornthwaite and al. : Geographical Rev., 38, 55 (1948)
5. C.W.Thornthwaite : Publications in Climatology, 17, 419 (1967)
6. H.L.Penman : Proc. Roy. Soc. 193, Londres (1948)
7. H.L.Penman : Noth. J. Agr. Sci. 1, 9 (1956)
8. M.Frere : FAO, Rome (1972)
9. C.B.Tanner, W.L.Pelton : J. Geoph. Res., 65, 3391 (1960)
10. P.Brochet, N.Gerbier : Ann. Agron. 23, I, 31 (1972)
11. P.Brochet, , N.Gerbier: Monographie de la Météorologie Nationale, 65, Paris, 67 (1975)
12. P.Seguin : Ann. Agron. 26, 671 (1975)
13. J.D.Horel : Mon. Wea. Rev. 109, 2080 (1981)
14. J.P. Piedelièvre, M.Deque, J. Servain : Notes de travail de l'EERM, 157, 58 (1986)
15. A.G. Barnston, R.E. Livezey : Mon. Wea. Rev. 115, 1083 (1987).
16. P.J.Lamb, R.A. Peppler : Bull. Am. Met. Soc., 68, 1218 (1987)
17. J.E.Namias: J. Geophys. Res., 85, 1585 (1980)
18. W.W.Kellogg and Zong-Ci Zhao : J. Climate, 1, 348 (1988)
19. Zong-Ci Zhao , W.W. Kellogg : J. Climate, 1, 367 (1988)
20. S. Manabe, R.J. Stouffer : J. Geophys. Res., 85, 5529 (1980)
21. S.Manabe, R.T. Wetherald , R.J. Stouffer : Clim. Change, 3, 347 (1981)
22. T.C.Yeth, R.T. Wetherald and S. Manabe : Mon. Wea. Rev. 112, 474 (1984)
23. S. Manabe, R.T. Wetherald : Science, 232, 626 (1986)
24. W.M.Washington, G.A. Meehl : J. Geophy. Res., 89, 9475 (1984)
25. S. Manabe, R.T. Wetherald : J. Atmos. Sci. 44, 1211 (1987)
26. G.A. Meehl, W.M. Washington : J. Atmos. Sci., 45, 1476 (1988)

Sustainable Green Energy, CO_2 Neutral, an Answer for Rich and Poor Alike?

C. Lincoln

Environmental Study for the J.W. McConnell Family Foundation,
2115 De la Montagne, Suite 200, Montreal, Quebec, H3G 1Z8, Canada

The gulf crisis finds the world fossil fuel dependent as never before. Meanwhile the other crisis, the environmental one, reaches critical proportions. CO_2 emissions must be stabilized, then significantly reduced. Sustainable energy through enzymatic conversion of biomass to ethanol is one ready CO_2 neutral answer both north and south should seriously consider.

There is no need to detail the countless cases and sectors of devastating evidence, backed by their formidable statistics. Suffice it to reiterate briefly that the technological developments of the twentieth century, impressive and creative as they have been - especially in the last fifty years - have caused ecological damage on a scale comparable, in the view of Professor Edward O. Wilson of Harvard, to that of the age of the dinosaur. Just the few statistics which follow tell their overwhelming tale. For one, the worldwide chemical production has jumped from 7 million tonnes in 1950 to 250 million tonnes only 35 years after, in 1985, an increase of 3,500%! It is forecast to double to 500 million tonnes by 1995. [1] Yet this considerable production, both in quantity and diversity, has been achieved without proper planning or security as to its consequences, not only in regard to the industrial processes involved or their effluents, but vis-à-vis the products themselves. The myriad problems engendered by pesticide products at all levels are a clear case in point.

No less dramatic than the rise in chemical production is the growth in car production. Outside the U.S.A., it has jumped from 7 million in 1950 to 125 million in 1980, an increase of 1,790% in 30 years. The world's motor vehicle population is reliably expected to reach or exceed 500 million by the year 2000. When it is considered that motor vehicle contribution to CO_2 emissions today attains 31% in the U.S.A., and 15% worldwide, the significance of the growth forecasts is just by itself an extremely alarming one. [2] [3]

Regardless of the terrible evidence of a planet on the brink of irreversible damage, effective remedial planning and action on a global scale has been spotty, uneven, and too often, uncoordinated. Whilst certain specific initiatives have enjoyed promising starts and represent hope for coordinated future action on global ecological issues (e.g., the 1987 Montreal Protocol on Substances that Deplete the Ozone Layer), there still seems to lack the determination within the international

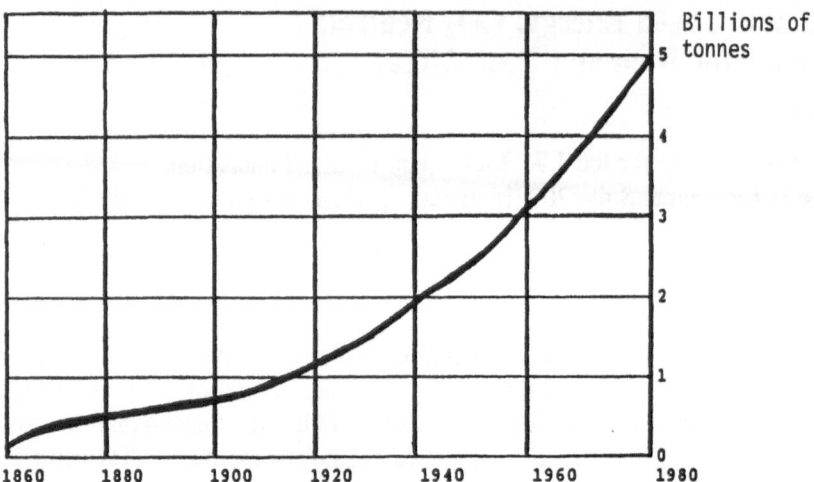

Fig.1. Global CO_2 emissions per year. Carbon dioxide emissions from fossil fuels have risen dramatically since World War II

community to view the need for a coordinated and comprehensive integrated strategy as a desperately urgent priority. It is almost as if, despite the overwhelming evidence, our governments and the peoples under their leadership consciously lead themselves to believe that the capacity of the planet to absorb punishment and renew itself is absolute and infinite.

Why look further than at the unfolding and striking example of the Gulf crisis? We suddenly woke up to the renewed realization that, despite the previous severe warnings, political and ecological, we were still as dependent on fossil fuel energy as we had been twenty years ago, when events, as well as scientists and environmentalists, combined to summon us to beware, and to change our ways.

And yet, evidence indicates that a significant part of the escalating damage to the biosphere can be traced to industrial man's decision to use fossil fuels for energy purposes. There comes an irreversible point to the capacity of the planet to absorb much more than the 500 billion tonnes of CO_2 released into the atmosphere over the last two centuries. See "Global CO_2 Emissions Per Year".

I have neither the authority nor the capacity to speak as a scientist, which I am not. However, I have read and heard and witnessed enough scientific and factual evidence over the last decade of political action to be increasingly convinced that it is high time for coordinated and integrated planning and action. The Gulf crisis may be the new catalyst for action that is yet another urgent signal to our leaders and ourselves. For within the global integrated action plan which must be achieved if we are to secure environmental peace, surely one of the focal components must be fossil fuels and the disastrous greenhouse consequences which their CO_2 releases are making more probable every day.

This said, I am keenly conscious that CO_2 releases are only one contributor to global warming, and that comprehensive solutions will depend on our taking into account linkages of causes, sources and consequences. However, the purpose of this paper is to deliberately focus on one specific issue of CO_2 build-up from fossil fuels, primarily in the motor vehicle sector.

The Gulf crisis was the main motivating factor in my selecting this narrower focus, as it brought home to me once again, and very forcibly, our massive dependence on oil. We in North America consume more than a quarter of world oil supplies, mostly to fuel our motor vehicles. Indeed, as the global consumption of motor-vehicle fuels derived from oil account for a significant portion of global CO_2 emissions, and thus of global warming, this paper attempts to highlight one CO_2 neutral substitute which could be made available within a realistic timeframe, especially for transportation fuel, and affordable by both North and South alike.

My experience in politics and government has taught me that it is impossible to rely only on market forces and voluntary constraints to redress major societal problems, and change behaviours and systems on a massive scale. For any strategy of massive overall change to become effective, there must be:

      (a)   leadership and direction towards a clear and understandable objective;

      (b)   definite timetables and targets to reach the overall objective.

To have any real value, the sentiments and tentative commitments expressed at Bergen, Norway, earlier this year of freezing CO_2 emissions at 1990 levels, must not remain wordy hopes but actively translate themselves on the home front into legislative and regulatory measures with targets, timetables, and teeth. Indeed, it is likely, when the latest scientific evidence is assessed, that we may have to agree on more stringent targets. See "Atmospheric CO_2 (PPM)". It is in fact suggested in authoritative circles that high-energy and medium-energy countries should reduce CO_2 emissions by 60% to 70% by the year 2030.

For this strategy to work, there will have to be measures taken, especially by the industrialized nations, on both the energy-demand side and the substitute-supply side. As a start, it is imperative that energy-wasting countries embark immediately on a massive program of energy conservation and alternative uses and processes of energy production and consumption. The world is particularly indebted to Amory Lovins and Rocky Mountain Institute for the measures and recommendations they have so courageously and steadfastly advocated over the years (long before these became fashionable!), and which Mr. Lovins will no doubt address at this conference.

Essential and effective as constriction of the energy demand becomes, the substitute-supply side is just as critical in reaching an overall solution to the CO_2 crisis. As the "problématique" of this conference has outlined so judiciously, any solutions must rest on equity, and ensure an opportunity for a balanced socio-economic quality of life for all nations, including emerging and developing countries. Many ideas have been advanced for CO_2 stabilization measures, including an overall energy distribution system, whereby ceilings would be placed on the various countries, beyond which ceilings "wasters" would have to contribute their energy rights to "savers" and "deservers". Even if such schemes were workable, they might tend to become an entrenchment for fossil fuels and CO_2

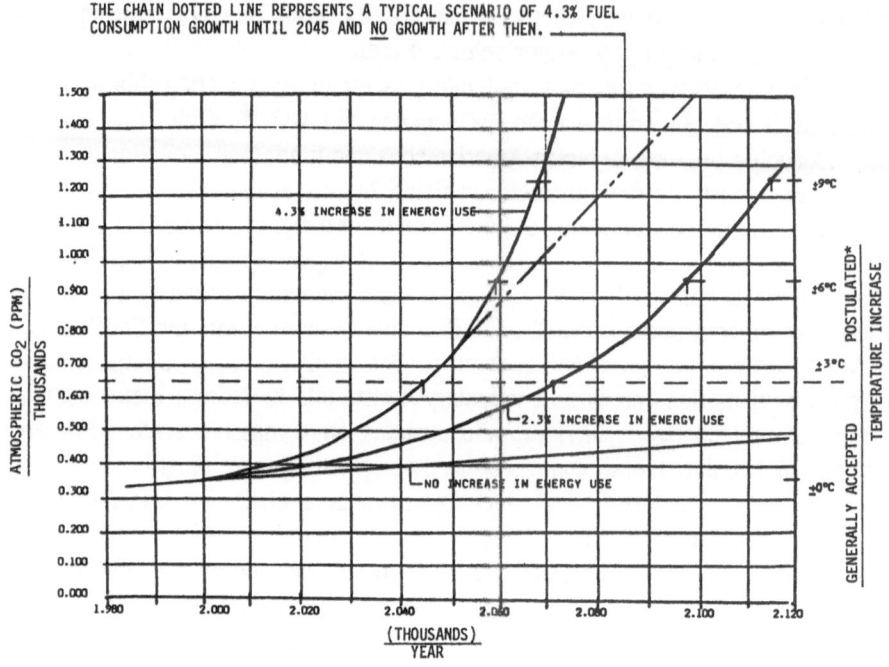

THE CHAIN DOTTED LINE REPRESENTS A TYPICAL SCENARIO OF 4.3% FUEL CONSUMPTION GROWTH UNTIL 2045 AND NO GROWTH AFTER THEN.

*PROJECTIONS ABOVE THE DASHED LINE ARE BASED UPON A LINEAR RELATIONSHIP BETWEEN ATMOSPHERIC CO_2 AND TEMPERATURE

Fig.2. Atmospheric CO_2

emissions, such being simply shifted from point "A" to point "B" to point "C". We have to look at CO_2 neutral fuels as the only effective basis for a successful substitute-supply side policy. Such commodities will become achievable and acceptable if they can be proven to be:

(i)   environmentally sound;
(ii)   safe, especially ecologically safe;
(iii)   effective and practical;
(iv)   available within a reasonable timeframe and accessible to developing countries and to other markets alike;
(v)   affordable.

As I mentioned, I am no scientific expert, my judgment being that of a committed environmentalist who has participated in legislative and executive politics, and in that context, been exposed to a great many proposals, projects and publications. Like all of you, I have attended countless meetings and conferences on the environment. I have been especially fortunate to have been a member of the Canadian delegation at the 1987 Montreal Protocol Conference, and of the National Task Force on Environment and Economy which followed the Brundtland Report. Like you, no doubt, I have ascertained that because of the variety and complexity, indeed the immensity, of the environmental problems and challenges, words and resolutions rush forward like white-water rapids - in stark contrast to

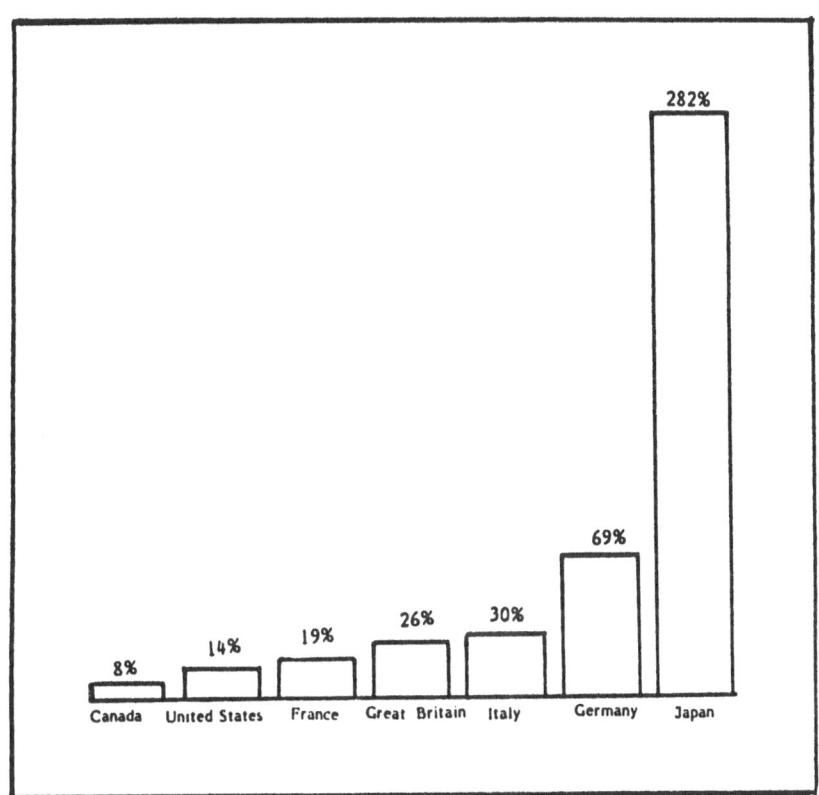

Fig.3. Percentage of farmland needed for complete gasoline replacement. A tree plantation 300 miles square (300 x 300) can produce all Canada's requirements for both gasoline and electrical power. A tree plantation 150 miles square can produce all Canada's gasoline requirements and all non-nuclear thermal power.

action, especially concerted action, which is naturally far more difficult and slower to achieve. This is why the lesson I have drawn - which may be simplistic, I concede - is to inform myself as conscientiously as possible on global issues, but narrow my action-perspective to one that is practical and achievable.

As a Canadian living in a country which is a renowned waster of energy; a country where investments in the energy sector constitute 1 in 7 total dollars invested; a country with 10% of its GNP in the energy sector [4] (about 85% of Canada's total energy comes from fossil fuels - about 66% of Canada's primary energy comes from the Province of Alberta -virtually all of that from fossil fuels) [5] - I have been particularly interested for several years in the key issue of substitutes for our overwhelming use of motor-vehicle fossil fuel.

Of what I have seen and heard, and examined, the fuel that has arrested my attention most steadily over the last decade is Sustainable Biomass Energy in the form of tree-derived (or comparable counterpart) ethanol. See "Percentage of Farmland Needed for Complete Gasoline Replacement" and "Greenhouse Effect - Energy From Wood is a Leading Supply Side Response to Global Warming". For

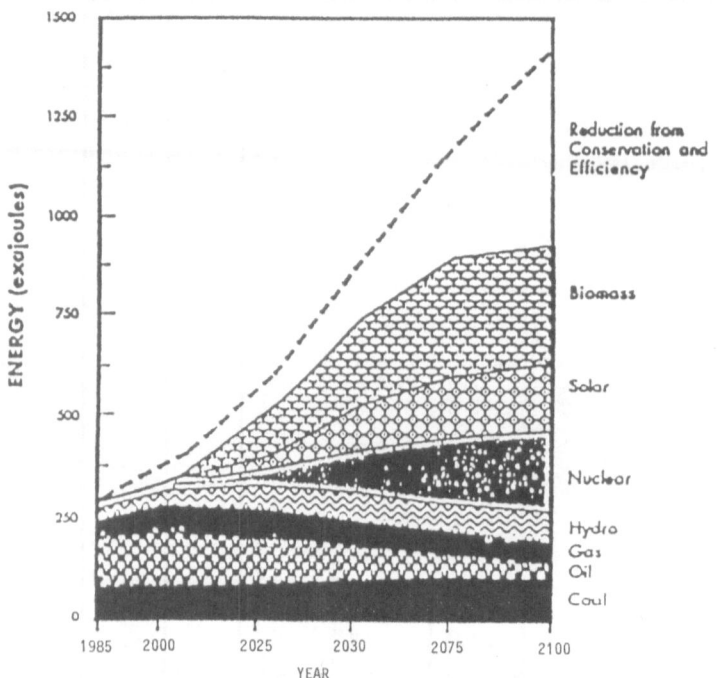

Fig.4. Greenhouse effect. Energy from wood is a leading supply-side
response to global warming. Primary energy supply by type. Source: US EPA
"Policy Options for Stabilizing Global Climate"

several years now, I have followed the evolution of sustainable biomass energy production from tree farming, and the subsequent (and successful) application of an enzymatic-conversion technology. In coming to the conclusion that this product and process are a prime candidate in our quest to provide an adequate transportation energy source as a CO_2 neutral substitute to fossil fuel gasoline, I have been motivated by the following factors:

1.   Tree farming and harvesting as carried out in the model I have studied [6] are ecologically sound and eminently feasible in most regions of the world, including the developing world. Moreover, they are sustainable over the long term.

2.   There is never one ideal solution. We have to look to all possibilities of achievable solutions, such as hydrogen with its considerable attractions as an effective and environmentally-clean fuel. What makes ethanol particularly interesting in my view is its practicality as a realistic potential substitute. The timeframes in which it can be put into production at an affordable cost and in significant quantities (as will be explained later) are reasonable, indeed promising. Not least important, its use as a combined fuel or as an autonomous fuel will not require any

change of significance to motor vehicles, and therefore to the production infrastructures of the automobile industry. [7]

3. Trees or biomass can be used for electricity, and/or be converted to ethanol, and are in fact more effective in this dual role because of the maximization of average yield per hectare farmed. However, as indicated, my particular interest has focused on tree-derived ethanol as a foreseeable substitute for gasoline.

4. Comparisons I have studied recently, especially since the latest oil crisis, indicate that the cost of ethanol is becoming increasingly competitive. See "Energy Produced per Acre".

5. The security advantages - including the key question of environmental security - loom extremely large. Both from the point of view of its far lesser volatility, and especially from the standpoint of marine spills, ethanol proves to be a considerably safer fuel (c.f. the disastrous aftermath of the Valdez spill, both in ecological and economic terms. The clean-up costs and liabilities, a poor substitute for a virgin environment, are evaluated at US $1.5 billion!).

It is forecast, with the support of a 15-year pilot project and sustained test pattern, both in the field and in a test enzymatic-conversion plant, that the following phases and targets are realistically achievable in Canada:

By 1995: Large-scale demonstration plant and first phase tree-planting by farmers.

By 2000: 2½% of current gasoline needs.

By 2015: 12% of current gasoline needs.

By 2030: 50% of current gasoline needs.[8]

In this connection, it should be noted that over 6.5% of all U.S. gasoline now contains fuel ethanol. This percentage climbs to over 20% in nine U.S. states and 30% in Nebraska. In fact, the U.S. consumption of fuel ethanol is as large as Canada's total gasoline market. [9]

It should be pointed out that the purpose of the demonstration plant is not to establish the success of the present enzymatic-conversion process, which has already been proven by the pilot plant. It is to fine tune the process on a large scale, and thereby reduce the product cost. The pilot project can now ensure a C. $0.35 a litre cost, which some laboratory experiments have been able to reduce to C. $0.24 a litre. C. $0.17 a litre is viewed as an optimum.[10]

Although the percentages indicated above refer to a Canadian context, the state of the process is such that the model could apply easily elsewhere - in fact more easily in many regions considering tree-growing climatic conditions.

I should explain at this stage that I am at present conducting a long-term independent environmental study for a large Canadian foundation. Neither it nor myself has any axe to grind. I have no profit or ulterior motive, my only motivation being the environmental cause to which I have been committed for

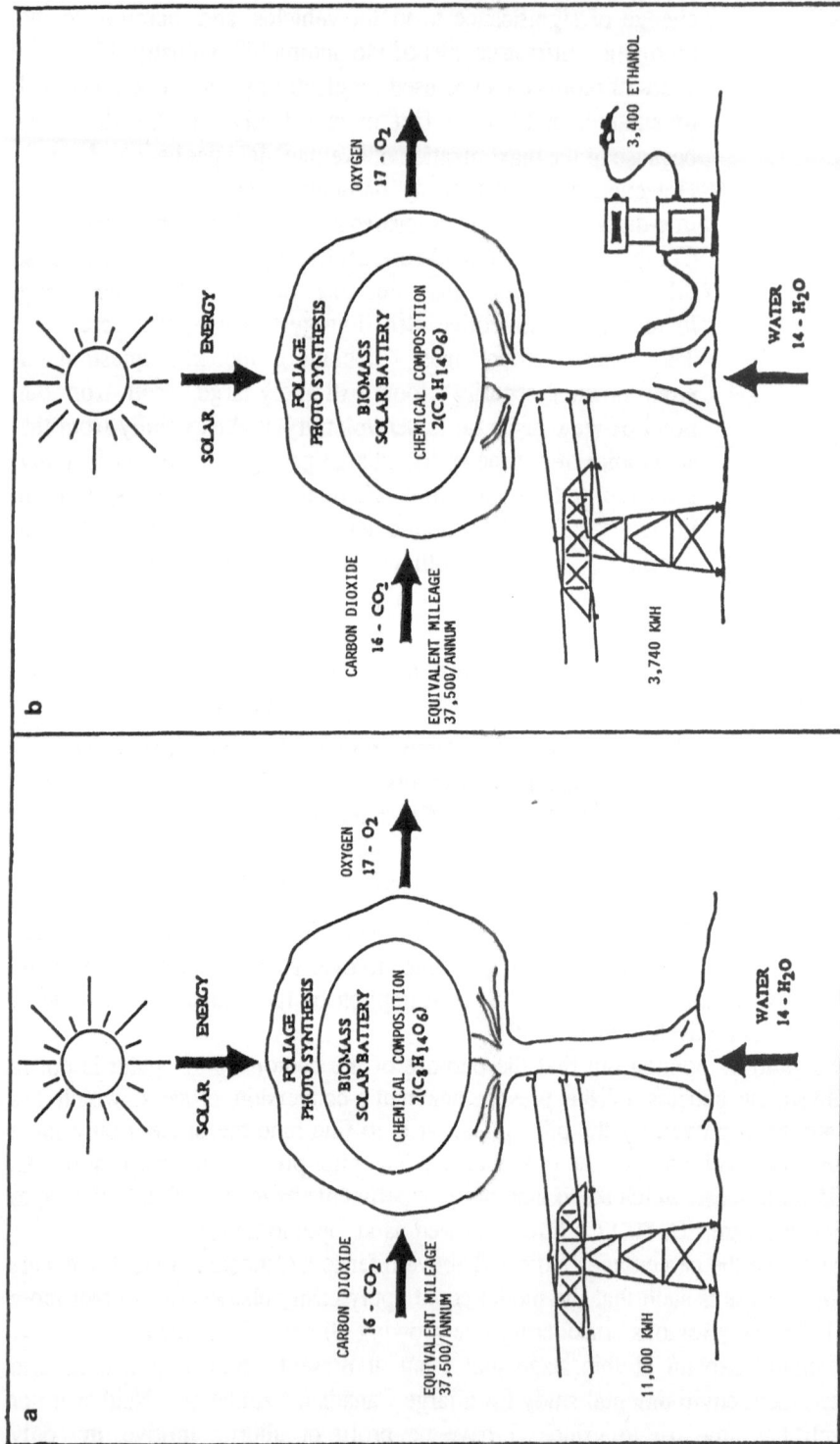

Fig.5. Energy produced per acre (yield 10 tonnes). (a) Electricity. (b) Electricity and liquid fuel

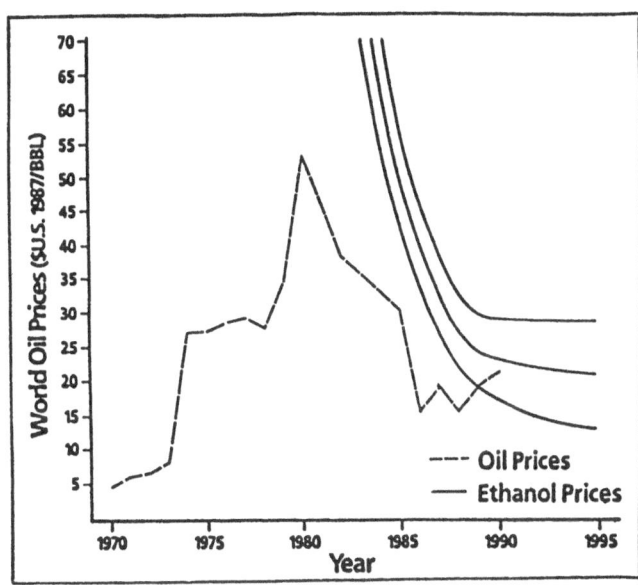

Fig.6. World oil prices vs ethanol from wood

several years, and to which all of you are committed. During my career in political life, as someone keenly concerned with environmental issues, I have watched my country - so blessed with bountiful natural resources and an immense territory - reinforce its dependence on fossil fuels and mega-energy projects. (I certainly do not intend to highlight my country for selective criticism, as all our countries have kept warm and continue to keep warm, under the cosy fossil-fuel blanket!). The latest such venture (for it is this, a venture) is the development of the Hibernia offshore oilfields at a cost of $5 billion, of which our government (we! ourselves!) will bear half. This new venture will provide fossil fuel at $38 per barrel or $24.96 "unburdened", not to mention the serious ecological risks inevitably involved, versus an "unburdened" cost for CO_2 neutral ethanol of $12.32 per barrel. See "World Oil Prices vs Ethanol from Wood". How can we claim to be attempting to reduce global warming in the first place, whilst forging ahead at huge expense - not just in Canada but in many other parts of the world - to drill the seas for CO_2-creating fuels? The logic of this double-entendre escapes me completely.

The Gulf Crisis has reinforced my faith in the conservation measures and alternative ways and sources long advocated by Amory Lovins and others of his school. Besides, it has also reinforced my conviction that we must force the pace of change on the oil companies and our governments by filling the supply side with practical and ecologically-sound substitutes.

I have hesitated long and hard before deciding to present a one-issue case before this august conference, convened to discuss the global challenge before us. If I have decided to do so at the risk of favouring what may appear at first glance the simple approach, it is that I am inherently an action-person, seeking practical results from worthwhile ideas which determination and better exposure can make possible. So if I make myself an advocate of sustainable biomass fuel today, it is because I have become convinced it is one CO_2 neutral alternative which can be realized, because it is relatively easy of production and of application, and increasingly affordable. Indeed, it may prove to become one of the prime fuels of the future for rich and poor alike, for the developing and the developed.

One cynic, but like me a convinced believer in sustainable biomass energy and the enzymatic-conversion process, commented to me the other day: "Maybe it is too simple and logical to excite the imagination of the powers that be!" Indeed! No wonder we have yet another crisis over oil, and are sinking billions into the sea to extract polluting fuels.

SOURCES

[1]   HAZARDOUS CHEMICALS. UNEP Environment Brief No. 4, 1986.

[2]   ATMOSPHERIC OZONE RESEARCH AND ITS POLICY IMPLICATIONS. Proceedings of the 3rd Dutch-U.S. International Symposium, Nijmegen, The Netherlands, May 9-13, 1988.

[3]   Francesca Lyman, with Irving Muitzen, Kathleen Courrier and James MacKenzie, "The Greenhouse Trap", World Resources Institute, 1990, Beacon Press.

[4]   Address by the Honourable Jake Epp, Minister of Energy, Mines and Resources of Canada, September 24, 1990.

[5]   Globe and Mail, Monday, October 8, 1990.

[6]   Brian Foody, "Ethanol from Biomass: the Factors Affecting its Commercial Feasibility", Ottawa, Canada, February 1990.

[7]   James R. Conrad, "Energy from Biomass - An Environmentally Friendly Future", an address to "Our World - The Summit on the Environment."

[8]   AGRO ENERGY, THE ECONOMICS AND THE FUTURE, Techtrol Ltd., Montreal, Canada, 1987.

[9]   James R. Conrad, N.7

[10]   BIO ENERGY, BIOMASS PRODUCTION, POWER GENERATION & CO-GENERATION, Iogen Corporation, Ottawa, Canada, and Techtrol Ltd., Montreal, Canada, February, 1990.

International Cooperation for "A Little Breathing Space"

M. Simai

Institute for World Economics, Hungarian Academy of Sciences,
1124 Budapest, Kallo Esperes 15, Hungary

Introduction

The problems of climate change have always been important issues. Human history is full of important events (sometimes tragic,) caused by climatic changes and catastrophes. In the past major migration waves have occurred as the result of long draughts, and in the late 20th century the potential for "man-made" climatic catastrophes has emerged as a grim consequence of the social and economic developments of recent decades. This is why international research programs on the human dimension of global change are increasingly addressing climate-related problems.

Climate impact studies have until recently been focused on the causes and potential consequences of the anticipated warming for countries and regions. Much less research has been commissioned on the cooperative regimes which may be required in future to deal with the problem. This paper raises some of the questions of international cooperation in a general way and also from the point of view of the Central and Eastern European region, which is one of the most heavily polluted parts of the industrialized world today.

The idea that excessive man-made emissions of carbon dioxide might disrupt the fragile balance of the atmosphere was raised first by a Swedish chemist in 1896, (Svante Arrhenius). Sixty years later, in 1958, the Scripps Institute of Oceanography in California indicated that half of the CO_2 released by industry was being trapped in the atmosphere and its level was increasing. Since then two major factors have become much clearer, and they have influenced thinking on ecological change. They are rapid world population increase and the interconnectedness of the factors influencing climate change.

World Population Trends

In the last part of the 20th century the relationship between human society and the ecological system has entered a new phase. The number of inhabitants on Earth surpassed 5 billion in the 1980s and by the end of this century it will be above 6.2 billion. (At the turn of the 19th century there were only about 1.6 billion people living on our globe.) The simultaneous presence of so many people on Earth has a major impact in itself on the ecosystem through human metabolism, production, consumption and deployment patterns.

The Global Environment
Editors: K. Takeuchi · M. Yoshino © Springer-Verlag Berlin, Heidelberg 1991

It is well known that the impact of human beings on the ecologically system increases faster than their total number. While the global population increased only by about 3.6 times between 1900 and 1990, global output increased more than fifty times, and energy output more than 30 times. Humankind is now using more than 40% of the total energy created by photosynthesis.

Global Interdependency

It is now better understood that ecological problems are interdependent, interrelated and international (and the ecological sciences are of an interdisciplinary nature). The global factors affecting the ecosystem are the simulative consequences of the integration of local causes and effects. At the same time they bring about new global effects which also influence local conditions. This international interconnectedness is a key factor in the debate about greenhouse gases and the process of global warming.

At the moment the data base is still not well developed. Disagreement among experts about the facts, sources, extent and consequences of global warming is due in part to the deficiency of existing data. Since the climate of the Earth has always undergone substantial change, some experts want more convincing proof that the warming observed during the past few decades has been caused by increased emissions of CO_2 and other greenhouse gases. Some think that what we may have seen has been a cyclical warming following the cooler period of the 15th to 19th centuries. Nobody is denying, of course, that during the last hundred years while there have been oscillations, there has been an overall increase in global mean temperatures. The hottest years in the entire period occurred during the 1980s.

There is also general agreement that the concentration of CO_2 in the atmosphere is higher today than at any time in the past 1-2 million years. Also it is generally agreed (even allowing for other sources) that man-made factors are playing a major role in the atmospheric concentration of CO_2 and of other greenhouse gases. There is, however, disagreement about the relative contribution from different sources.

Varying Contribution of Regions and Countries

While the global character of most ecological problems has been generally recognized there are important differences in the role of particular countries and regions. The causes of damage and the specific vulnerability of certain regions vary. Some experts conclude from this fact that instead of trying to create global cooperation, those countries which are most affected should act, individually or jointly, to address ecological problems. Other experts believe that a large part of the northern belt of the Earth may even be favorably influenced by global warming.

The potentially disastrous effects of global warming may of course be concentrated in mid latitudes and in particular coastal areas. However such consequences as the impact of changing rainfall patterns on forests, deteriorating food and fresh water supplies in certain regions, as well as floods, may lead to broader ecological effects and mass migrations. Climatic extremes may cause global tragedies. The potential consequences of the greenhouse effect are therefore global, and without global cooperation it will not be possible to create the necessary conditions for reversing causes, and developing effective responses. One cannot and should not deny that fact either, that there are special national and regional interests, which influence the "degree of propensity" for international cooperation.

The ecological systems of the world may be seen as a global bank from which all nations borrow in order to "finance" their individual economic development. The nations of the world are at the same time its stockholders. Its assets are the earth natural resources, the air, the water, the flora and fauna, the minerals. In the global environmental bank, the Soviet Union and Central Eastern European countries have been heavy borrowers in the past 40 years and they have deposited disproportionally large quantities of waste, in relation to their share in global output and consumption. Their contribution to CO_2 emission, for example, is 26%, which is relatively high in comparison to their 17% share in global output. As they are located in the temperate climatic zone, their water supplies and climatic conditions may be radically modified by CO_2 and other greenhouse gases. As has been seen from the experiences gained so far in international environmental cooperation, national and international regimes are closely interrelated.

The Role of Developing Countries

The poorer a country is, the more limited is its capacity to address ecological damage, or to develop ecologically sustainable systems, or to respond to climate changes. Some countries such as India and China have had experience over the past centuries in dealing with flood and drought. Today, however, the problems are deeper and broader. Increasing environmental problems are the cause of the fact that the wealthier countries are becoming richer, and the poor countries poorer, even in the field of ecology. Poor countries usually burn fuel less efficiently and technology which is more polluting. They sometimes "receive" more polluting industries, when it comes to global redeployment by multinational companies. The shrinking of the tropical forest resource is also caused to a great extent by poverty, and the nature of land tenure.

In the condition of poverty the priorities of a population are influenced by employment and other basic needs considerations, and not by such issues as the role of the environment in shaping quality of life. The educational level of the population of course influences ecological consciousness in most countries. This includes not only an understanding of ecological problems, but also a readiness to participate in efforts to protect the environment and work for sound national and international environmental policies.

Poorer countries are of course much less able to implement
the kind of environmental measures envisaged by international
agreements. Some developing countries already have very large
populations and world population growth during the coming
decades is likely to be concentrated in these same areas. The
impact of energy use and of further deforestation in these
countries may have further devastating global consequences.
Besides the general humanitarian arguments this is one of the
reasons why the more affluent countries should help the poorer
ones to implement internationally agreed environmental
policies.

International negotiations on the reduction of greenhouse
gases have not yet started, but there is an increasing
consensus among the experts of industrialized countries that
the developing countries will require special assistance in
their energy policies and in their efforts at adjustment to
climatic changes. One major industrial country, Japan, has
already declared such a commitment. In 1989 Japan expressed a
readiness to provide US$2.25 billion over the next three years
in environmental assistance, and the Japanese Environment
Agency has started working with other countries in the region
to develop an Asia-Pacific program to respond to climate
change.

The Situation in Central and Eastern Europe

In the Central and Eastern European countries past regimes -
as has already been mentioned have been by and large
inefficient in dealing with environmental problems. There
were of course differences between the more liberal Hungary
and the more rigid Czechoslovakia and GDR regimes, in
evaluating and managing certain ecological problems. No
Central and Eastern European country has had, however, a
comprehensive and realistic environmental strategy, and there
has been no regional strategy within the framework of the CMEA
to take account of the conditions of the ecosystem and the
impact of change. As a result of their inefficient energy
use, their contribution to the global greenhouse effect is
greater than their share in global production.

Over the last 40 years East Central European countries and
the USSR consumed about 30-40% more energy and raw materials
per unit of output, than the Western industrial countries or
the industrializing developing countries. While during the
past 15 years there has been some decline in the material and
energy consumption of the GDR, this has been far below the
trends of the Western countries. On average the six countries
of Eastern Europe still use more than twice as much energy per
unit of GDR as the more industrialized countries of Western
Europe (The Economist, Feb. 17, 1990).

There are several causes of this situation, but most are
connected with the functioning of a centrally planned system
and with the assumptions on which the system has been based.
Historically, the over-emphasis of the quantitative approach
and the neglect of quality and efficiency implied a reliance
on the gross value of production, in the evaluation of
performance. Such policies did not stimulate the conservation
of materials and energy, since in the final analysis this
would have slowed down the artificially inflated growth rates.

Most remuneration was closely tied to growth rates of gross output. Thus the whole system from workshops to planning agencies stimulated the use of more material inputs, because of the approach to evaluating performance.

In the USSR, the most important commodity producer and the main supplier of raw materials in the region, natural resources were in general treated as "semi-free" goods. They were intentionally underpriced, production costs did not include such factors as the costs of the infrastructure. The patterns of industrialization, the over-emphasis on heavy industry, the neglect of the service sector, which was also partially ideologically motivated, resulted in an extremely heavy material and energy intensive economy. Due to the relatively low commodity prices, there was no interest in investing in technology which would bring about major economy in energy or raw materials. There was very little research and development in this field. For decades the member countries of the CMEA have traded energy and raw materials with the USSR in return for their products and this also provided an important incentive for waste and over-consumption.

The price explosion during the 1970s did not lead to any major improvement in the use of energy or raw materials per unit of output. Some decline took place in certain countries, but the basic tendencies did not change. The whole economic and especially the industrial system placed heavy pressures on resource production in the region. These policies lead to a much faster exhaustion of supplies, and later a major shift in investment from manufacturing and services to energy and raw material production, due to the fact that for example in the USSR, the raw material resources were located in remote areas, with no infrastructure.

The Environmental Consequences of Economic Policies

The consequences of the wasteful use of resources has had a damaging effect on the environment, and the present environmental crisis in large areas of Czechoslovakia, Hungary, Poland and the GDR is the result of these practices. One example is the pattern of fuel use. The main fuel in Central Eastern Europe is coal, especially brown coal or lignite, which has high sulphur and ash content, coupled with low heating efficiency. The cars produced in the region burn more gasoline and their polluting effect is higher than in the West. The inefficient and wasteful use of fertilizers has been another important source of environmental damage. In Czechoslovakia, for example, the use of fertilizers increased twenty times during the past three decades, but the yields in agriculture only doubled, 54% of the arable land is endangered by erosion (Die Weltwoche, 1989. No. 22, p.5.)

For decades the authorities neglected environmental problems and the relationship between economic growth and the ecological system. The five year plans did not deal explicitly with environmental issues up till the 1970s. Even after the 1970s it was assumed that governments can efficiently harmonize economic and social development with ecological change. A belief in the unlimited domination of "man over Nature" pervaded the thinking of the planners. Environmental damage was often kept secret. An internal confidential decree in 1982 in Czechoslovakia, for example, refused to permit the publication of any information concerning the real situation in the ecosystem. This official attitude was similar to that in other countries in the region. Mostly it was academics and opposition groups who started to publicize, often illegally, the environmental crisis.

A typical example of this approach was the designing of the Bos-Nagymaros hydroelectric complex, a joint Czechoslovakian-Hungarian project, which almost completely disregarded the environmental implications and consequences of the project for the two countries. Because of the monopolistic nature of the ruling parties, it was very difficult to build up strong, popular grassroot movements to impose pressure on governments. There were environmentalist NGOs in every country, within the framework of the "official" social movements. There were also opposition groups which in certain cases played an important role, as in the struggle against the construction of the Bos-Nagymaros hydroelectric project in Hungary already mentioned.

Post 1970s Environmental Policies

An institutional structure to begin to deal with environmental problems emerged after the 1970s in each country in the region. (In the academic community, the concern about problems of pollution and the wasteful use of raw materials and energy, of course arose much earlier. The political leadership, however, very seldom listened to the academics and they were sometimes accused of being under the influence of Western propaganda, and of "over-emphasizing" environmental problems.) Unfortunately the institutions which became responsible for environmental matters also held other conflicting responsibilities.

Since the late 1970s parliaments have adopted a number of laws on environmental and management policies which were supposed to be the basis of central government decisions relating to economic development and environmental impacts. Parliaments adopted five-year plans which included the allocation of natural resources. However, in general, environmental policies continued to receive low priority in the national plans.

Different government agencies were charged with carrying out economic and legal measure for environmental protection. The same state organs were however often responsible, in the centrally planned system, for the establishment and the operation of those economic activities which used raw materials and energy and which were the fundamental sources of pollution.

Even when these opposing responsibilities did not coincide, it was the central government agencies which set the standards for tolerable pollution levels and pollution control. They decided the sites for new extracting industries, new manufacturing services and new settlements. They determined also the patterns of output and tried to influence the structure of consumption, the use of natural resources. It was also the state which decided what legal sanctions should be imposed on environmental violations.

Because of this distribution of responsibilities it was natural that no central government agency ever closed down any large state-owned factory for environmental reasons, since the output of the particular firm had been a source of revenue in the national plan determined by the same government. Even in those countries where economic reforms were introduced, state enterprises operated in a hierarchical institutional system. The responsibilities of the enterprise managers and the central bureaucracy could not be separated and this had an adverse influence on environmental policies and practices.

In principle local authorities were supposed to play an important independent role in environmental management, even in this system. However, while the role of local authorities was not the same in all the socialist countries, in many cases their tasks were confined to the implementation of central political and economic decisions. In a few countries they had a limited opportunity to protect local environmental interests and they became more active in environmental matters.

The record of the agencies dealing with the ecological problems in Central and Eastern Europe is far from satisfactory. It is true that these agencies have not been too successful in the Western market economies either, but at least they could deal with certain problems more efficiently. They have had some successes in such areas as reducing urban pollution, promoting river rehabilitation and monitoring damage. In Central and Eastern Europe, even where they were independent, the environmental agencies held a weak position in the political hierarchy. They were powerless against strong lobbies, and also had major fiscal constraints. Hence the agencies working against pollution by firms or institutions were by and large inefficient.

Generally, low fines were levied for violating environmental legislation, and when - in exceptional cases - they were higher, they failed to influence management, because they were paid from operating costs and passed on to consumers. In Poland a survey of pollution fees paid by 1400 manufacturing enterprises found that on average these amounted 0.6% of the production costs (The Economist, Feb. 17, 1990). The monitoring and inspection agencies were in general not only understaffed, but had relatively few trained experts, and did not have efficient enough instrumentation to measure impacts or determine sources of pollution.

The governments of the Central and Eastern European countries became partners to many international and regional agreements dealing with ecological problems, especially environmental conservation. In many cases, however, they were not able to fulfill many of the obligations stipulated by the agreements. There was a conflict between national and international standards, and the ability and readiness of the Central and Eastern European countries to implement the agreed measures was lacking.

The ecological crisis was one expression of the fact that, within the framework of the given politico-economic system, countries seriously violated such environmental limits as the availability of natural resources, and overloaded air, water and land with waste and pollutants.

The Consequences of Political Change

The transition from a centrally planned system to a market economy and from a one party rule to a democratic, pluralistic multi-party system cannot be achieved simply by popular votes or by government decisions. It will take a very long time. Just one issue: to build up a market-led economy requires profound social changes, and the development of and entrepreneurial mentality among the population. In macroeconomic management it is necessary to have a realistic price system, and proper exchange and interest rates which reflect the real cost of money. As far as environmental protection is concerned, the cost of effective conservation measures is a difficult problem even in well functioning market systems.

In addition it is still an open question what direction the changes in economic development will take, in the countries of East and Central Europe. It is evident from the experience of the industrially developed countries, without deliberate policies, that the market is not a much more efficient instrument than central state planning in promoting sustainable development. Certainly, the market-led economies reacted to price increases in raw materials and energy during the early 1970s by stimulating the rapid advance of miniaturization, the introduction of new materials and components, recycling and other measures. This was a more efficient response than that of the Central and Eastern European countries.

Taking into consideration the grave economic and social problems of the Central and Eastern European countries, the level of their international indebtedness, (the amount of which is close to US$100 billion, without the debt of the USSR,) the fact that the budgets of the central governments are oversized and must be reduced everywhere, that there is pressure to improve people´s standard of living, it will be very difficult for some years to expect major spending by central governments on environmental conservation. Resources for environmental conservation and rehabilitation must come from a variety of sources beyond the central budget, from local sources, from enterprises, and from grassroot actions. International assistance may also need to be an important factor in funding, organization, and education in environmental management.

Because of their domestic economic crises and limited resources, the Central and Eastern European countries need international assistance in some of these important areas:

- The training of experts;
- The introduction of expertise by the use of foreign specialists;
- The introduction of new technology to cause less or no damage, or to rehabilitate the damage caused so far;
- The radical modernization of the energy sector;
- The development of an effective monitoring system;
- The development of sound environmental policies, under the conditions of the market-led system.

Calculations have been made in some Western European countries showing that it would be more efficient for them to invest in the environmental rehabilitation and protection of Central and Eastern Europe than in their own countries, due to the wide regional impact of the environmental crisis in Central and Eastern Europe.

The World Energy Economy

In discussing the causes of global warming the impact of the world energy economy has been considered as a central factor. Its "greening" has become an important aspect of environmental policies. This, however, is a very difficult issue. There are great inequalities between countries in the use of energy. Most of the worlds energy production and consumption takes place in the industrialized countries, which have cumulatively contributed to the bulk of excess CO_2 in the atmosphere. There are great differences between nations in the energy requirements for economic growth. Especially in the poorer countries the energy needs and the improvement of the energy economy require large investments.

Systemic changes in Central and Eastern Europe cannot and will not eliminate rapidly those damaging environmental problems which are connected with structural backwardness, or with the predominance of material and energy intensive industries and obsolete technology. It will be difficult and expensive, for example, to improve coal-burning technologies and implement other measures which diminish emissions of CO_2 and sulphuric compounds.

There is also competition for energy among the world's economies and most governments are involved, because of economic needs and the strategic importance of the energy sector in growth and competitiveness. I quote the views of an international business research institute. "The complexity of the negotiation process will increase as the number of stakeholder nations grows. Any meaningful agreement on global warming, for example, will require the participation of a sizeable majority of industrialized and developing nations. There is clearly a need for countries to tackle the problem of global warming, an environmental issue for which the scientific community has shown almost universal concern. Unilateral action, however, does not guarantee environmental success. In fact, a country that would unilaterally impose stringent regulations or heavy carbon taxes to reduce CO_2 emissions, the principal greenhouse gas causing global warming, risks slowing its economic growth while achieving only marginal environmental improvement. Production costs could increase within that country, leading some energy-intensive industries (and accompanying CO_2 emissions) to simply relocate in more permissive jurisdictions. The benefits of concerted action are therefore clear. (International Business Research Center, The Conference Board of Canada, "Global Business Issues", September, 1990.)"

The world energy economy and its role in global warming has created great interest in the section of the scientific community dealing with the greenhouse effect.

Strategies for Global Cooperation

A very important scientific contribution to the debate on global warming and international actions required to arrest it, occurred in Budapest in April 1989, at a workshop sponsored jointly by several important scientific organizations, participants in a project on "The Human Dimension of Global Warming". The official title of the workshop was "A Little Breathing Space", chosen to indicate its concentration on the problems of the atmosphere.

The Budapest workshop, involving the participation of more than 80 experts from all parts of the world, devoted most of its deliberations to the proposals of the Toronto conference, held in June 1988, which recommended that global CO_2 emissions should be reduced by 20% until 2005, and by at least 50% until 2050. The Budapest workshop suggested a broad agenda for further research on different aspects of global warming. Special emphasis was given to further research on global cooperative regimes.

The Budapest workshop emphasized the following problems:

(a) The targets cannot be applied across the board to all nations.

(b) The Toronto target was not based on any optimization strategy, or on an analysis of what would be needed to avoid specific adverse effects.

(c) The role of other gases contributing to the greenhouse effect is rapidly increasing.

The 20% target, as a first step, was of course endorsed. In the suggested research agenda a number of important issues were raised in connection with international institutional implications.

Firstly, the workshop questioned the justification for a cooperative regime to deal with the greenhouse effect, and the feasibility of such a regime, taking into account the diversity of interests, and the competition in the sectors responsible for most of the emissions. Also it questioned how such regimes should be designed. It also questioned how an international regime could suggest appropriate national regimes and laws and bring about cooperative research.

An official answer to one of these questions may come relatively soon. The World Meteorological Organization and the United Nations Environment Program are in the process of setting up negotiations for a global convention on climate change. These negotiations could be extremely important for the future of global ecological and economic systems. The official aim of the negotiations is to discuss climate change and the protection of the atmosphere. Due to the nature of the problems they will have to raise fundamental issues of national energy policies and the structure, technology and other practices of the energy sector. They will have to deal with the problems of the world forest economy and with certain aspects of agricultural and food production. The outcome of the negotiations will also influence global and national policies on urban settlements.

All these and perhaps other problems imply that the cooperative regime which is necessary for global warming cannot be the task on one, existing or new, international organization only. A network or organizations is required for the implementation of an internationally agreed program of action which might be drawn up following negotiations.

Regional organizations within the framework of the UN and other regional groups will also have to play a crucial role, because of the particular regional impacts of global warming.

Global and regional actions must of course be based on national policy measures. These must include the elaboration and implementation of new energy policies and energy efficient economic policies, the preservation and the sustainable use of forests, and appropriate agricultural measures.

As far as research requirements are concerned it is necessary, if there is to be a positive outcome to negotiations, to intensify international comparative research in areas such as the following: measurement of climate change, modelling and projection of future climate change, better understanding of the impact of climate change and appropriate adaptation policies, (including their human, institutional, financial and technological needs.) More research is needed also on the comparative efficiency of different energy policies, both in terms of alternative sources of energy and the comparative analysis of structural policies causing energy savings.

All these measures will require extraordinary commitment and cooperation from governments, international organizations, private companies and transnational corporations. Strong popular support will be indispensable. A much higher level of partnership between all countries but especially between developed and developing countries is a fundamental condition of any successful international effort, if we are to reduce existing political, economic and institutional barriers.

In the evolving global political environment the necessary level of cooperation, within the context of a global strategy, may become a feasible and realistic goal.

Further Reading

Fischoff, Baruch Hot Air: The psychology of CO_2 induced climatic change. In Cognition, Social Behavior and the Environment, ed. J. Harvey, 1981. Hillside N. J., USA. Erlbaum.

International Conference of Global Warming and Climatic Change. New Delhi, 1989. Tata Energy Research Institute.

Jodha N. S., Potential Strategies for Adapting to Greenhouse Warming: Abatement and Adaptation. Washington D.C., 1988. Resources for the Future.

Sustainable Development and Science Policy. The Conference Report. Oslo, August 1990. N.A.V.F.

Speth, J. G. The Environment. The Greening of Technology. In the Journal Development, 1989 No. 2-3 pp. 30-32.

Understanding Global Environmental Change. Roger E. Kasperson, Dow, Golding. Ed. Report on an International Workshop. Clark University, 1989. ET-90-01. HDCP-RA-R 001.

The Role of the Ocean in Climate and Climate Change

R.W. Stewart

4249 Thornhill Crescent, Victoria, B.C., V8L 3G6, Canada

The ocean plays a key role in the climate system of the earth. Its characteristics are so very different from those of the land that without it the earth´s climate would be quite unlike the one we know. Because of the ocean, temperature ranges are very much reduced, particularly in regions not too remote from the coast. The large effective thermal capacity of the ocean causes the seasons to be delayed with respect to solar heating cycle. Because of ocean currents, some regions are warmer and others colder than they would otherwise be.

Some aspects of ocean circulation have characteristic times of centuries, and so act as a flywheel on the climate system, delaying response to changes in driving forces.

Also the ocean contains a very large reservoir of carbon dioxide and related chemicals. This reservoir exchanges with the atmosphere, but most of it resides in the parts of the ocean which have long time constants and so does not quickly come into equilibrium with the atmosphere. In addition, these carbon compounds are importantly involved in the oceanic biological cycles, which not only influence the distribution among particular carbon compounds, but strongly influence the distribution in space.

Although we understand these characteristics in broad outline, there remain many serious deficiencies in our understanding. Major projects have been mounted, under the World Climate Research Program and the International Geosphere Biosphere Program to reduce these deficiencies.

Introduction

Concerns over "Global Change" are attracting increasing public attention. High among these concerns is the question of global warming, arising from the known inexorable increase in the concentration of such greenhouse gases as carbon dioxide, chloro-fluoro-carbons, methane and nitrous oxide, together with the results of running large numerical meteorological models. These models, which do a fairly good job of reproducing present climate, indicate that if the greenhouse gases continue to increase in the atmosphere at rates consistent with expected economic growth, the mean surface temperature of the earth must be expected to rise by several degrees in less than a century. Such a rate of rise would be unprecedented, at least for the past 10,000 years, and would lead to temperatures which have not been experienced on earth for a million years or more. There is fear that the earth´s

The Global Environment
Editors: K. Takeuchi · M. Yoshino © Springer-Verlag Berlin, Heidelberg 1991

ecology, including humans might not be able to adapt to such rapid change without great disruption.

One of the questions with respect to global warming is the fact that these same models indicate that the earth should already be about one or two degrees warmer, because of additional greenhouse gases already in the atmosphere, than it was a century ago. Observations, although not entirely unambiguous, suggest that any temperature rise over the last century has been more nearly one half degree. However, it is known that the ocean plays a very important role in determining the details of today's climate. It is also known that the ocean is not well treated in existing numerical models. Therefore, increasing attention is being paid to the role of the ocean, to determine, for example, whether it merely delays possible effects or whether it importantly modifies them.

In addition, it is known that the rate of increase of carbon dioxide in the atmosphere is only about half that which might be expected from the known rates of burning of fossil fuels and the estimated effects of changes in the biosphere - notably the clearing of tropical forests. The fate of this "missing carbon" remains uncertain, but the ocean is a very likely candidate for its absorption, particularly since the ocean already contains very much more carbon than does the atmosphere.

In the ocean, the biota play a central role in determining the distribution of carbon and its interaction with the atmosphere.

These questions are now being actively studied in large international observational and modelling studies, notably the World Ocean Circulation Experiment and the Joint Global Ocean Fluxes Study. However, it will be at least a decade before they are adequately resolved.

In addition to these problems, with a time scale of many decades, the ocean plays a central role in determining the nature of inter-annual and inter-decadal climate fluctuations. However, these questions, although very important, are beyond the scope of the present article.

How the Ocean Influences Climate

Most of the effect of the ocean on climate can be understood in principle by considering its most fundamental properties: The ocean is wet. It is fluid. It is salty. It is deep. It extends from tropical regions into high latitudes. It is divided by north-south continental barriers. It is water.

Let us consider these properties as they affect the influence of the ocean on climate.

The Ocean is Wet

In a micrometeorological sense, being wet means that under most circumstances it loses heat because of evaporation, and that this heat loss increases rapidly as the temperature rises. However, unlike the case for most land surfaces, no

matter how much evaporation takes place, the surface remains
wet. The sun is able to provide only so much energy per unit
area. As a result, the surface temperature of the ocean is
limited by evaporation. In practice this limit turns out to
be 30°C, although we don't know exactly why it proves to be
this particular figure. Over most land surfaces, the
temperature can become substantially higher.

The Ocean is Fluid

Because the ocean is fluid it is readily able to move. One of
the important immediate effects of this property is the
behavior when the surface is cooled. The cooled surface water
becomes more dense than water just below, convection sets in
and there is a vertical mixing process which carries heat (and
other properties) toward the surface and surface properties
into the interior. Overnight cooling rarely amounts to much
more than 1 , and the convection extends to a few meters.
Winter cooling can be up to about 10° and typically extends to
a depth of the order of 100 m. Thus the ocean stores heat
during periods of strong insolation and releases it when the
sun is low or absent.
    While typically the depth to which convection penetrates is
a small fraction of the ocean depth, occasionally and locally
the surface cooling is so intense, and other conditions are
suitable, so that it continues right to the bottom, When the
bottom is reached, there is no more warm water to mix in so
that the whole water column, from top to bottom, can cool
appreciably. This column becomes heavier than surrounding
water and will eventually sink and spread out over the bottom.
The process is called "bottom water formation", and occurs
importantly in the Antarctic, high northern latitudes of the
Atlantic and in the Mediterranean Sea.
    Over land, on the other hand, there is no convection and
heat passes into deeper layers only by the slow and
inefficient process of conduction. Land surfaces therefore
both warm and cool much more rapidly than does the ocean
surface, so that there can be much larger day-night
temperature differences and much larger winter-summer
differences. Further, the heat storage capability of the
ocean retards the phase of the annual cycle relative to that
of the insolation so that typically the ocean surface is
warmest around the autumnal equinox and coolest around the
vernal equinox. (This is the reason why in many cultures the
seasons are defined as times delayed about six weeks with
respect to insolation.) Over land far from large bodies of
water, the seasonal temperature cycle much more closely
follows that of insolation.
    These resulting differences in amplitude and in phase of
the temperature cycles between land and ocean have very
significant effects on the atmosphere and on climate. Thus
maritime climates tend to be "equitable" compared
with"extreme" continental climates. Diurnal land-sea breeze
cycles, and annual monsoon cycles develop because of the
contrasts between the land and the sea. Air rises over the
warm area, sinks over the cool one and these is a surface flow
from the cool towards the warm.

So far we have discussed vertical water movements. The ocean, of course, also moves horizontally, carrying its properties with it. These horizontal movements also have very important climatological effects, which will be discussed later.

The Ocean is Salty

Ocean water contains about 3.5% dissolved salts. This proportion is sufficient to mean that it has no temperature of maximum density above the freezing point, analogous to the 4° C situation of fresh water. Also, the freezing point is depressed to about -2° C. As a result, the ocean resists freezing compared with freshwater lakes, even deep ones. When water freezes, the surface is able to become much colder, and heat transfer to the atmosphere is reduced. Thus ocean water is more effective in warming the air under cold conditions than is fresh water.

Perhaps more important, salinity is not uniform and more saline water is more dense than is less saline water. Over most of the ocean, salinity varies by only a few parts per thousand. If the water is warm, say above 10° C, the density is largely determined by temperature since ordinary temperature changes of a few degrees cause larger density differences than do ordinary salinity differences. However, near the freezing point, density is very insensitive to temperature and so is almost entirely controlled by salinity. As we shall see, this has very important consequences at high latitudes.

The Ocean is Deep

The average depth of the ocean is about 4 km. This far exceeds the 100 m or less for which there is appreciable penetration of sunlight. Thus deep cold water formed in regions of deep convection cannot be warmed by the sun unless some ocean circulation brings it close to the surface. It can only be warmed by mixing with warmer water from above. This mixing turns out to be very slow and inefficient, so the warming process can take as long as about 1,000 years. (Geothermal heating from below turns out to be negligible in this process.)

The general picture for the deep circulation is as follows: Surface water is cooled in the winter by negative radiation balance, evaporation, and direct heat transfer to the atmosphere. In addition, the salinity may be increased by excess evaporation over precipitation or by the expulsion of enriched brine from ice. If the temperature becomes low enough and the salinity high enough, bottom water formation takes place. This bottom water spreads out into other parts of the ocean. It is gradually heated, and thus made somewhat less dense, by mixing with slightly warmer water above. At the same time, newly formed bottom water slides under it, pushing it up from below. It seems that on the average enough new bottom water is formed each year to push other water up by about 4 m. Thus it takes about 1,000 years for it to be pushed up close to the surface. There, it becomes caught up in the surface, wind-driven circulations and eventually is carried to high latitudes where it may again become involved in a bottom-water-formation process.

One thousand years is a long time! During the past thousand years, there have been quite significant fluctuations in climate, including the medieval warm period and the little ice age. Therefore there is little reason to believe that the present ocean is in a steady state. Indeed it is quite improbable that the present climate, if continued over 1,000 years, would give rise to the present ocean. In particular, it seems likely that water in the present deep North Pacific is a relic of a previous climate. Similarly, if our future climate changes significantly, the ocean will not immediately come into equilibrium with it but will retain characteristics determined by present and past climates.

The Ocean Extends From Tropical Regions Into High Latitudes

Warm tropical surface water evaporates freely. Much of the water evaporated falls as tropical rain, but an appreciable proportion is carried to higher latitudes by the atmosphere. Thus the ocean surface becomes enriched in salt. Ocean currents are able to carry some of this high salinity water to high latitudes where it is cooled. If it retains its high salinity, it can then be available for deep water formation.

The Ocean is Divided by North-South Continental Barriers

As a result of the accident of continental drift, at the present epoch in geological time, major continents extend north-south, unbroken, for great distances, particularly in the northern hemisphere. This has several consequences. One is that it permits the formation of swift western boundary currents such as the Gulf Stream, the Kuroshio and the Oyashio, which carry water north-south fast enough that it retains its properties. Such currents significantly modify the climate in nearby areas relative to other areas at the same latitude.

Another is that, in particular, the North Atlantic and the North Pacific are separated and are free to behave independently of one another. In the North Atlantic, warm saline water is transported by surface currents through the passages between Scotland and Iceland. There it gives up its heat to the atmosphere, producing the mild European winters so characteristically different from other areas of similar latitude. Bottom water formation takes place, making room for more warm surface water to move in. In the North Pacific, the situation is quite different. The shape of the continents prevents the water from moving so far north. Excess precipitation over evaporation blankets the area with relatively low salinity water, and deep water formation is impossible. Warm saline water carried north by the Kuroshio must recirculate southward on or near the surface.

The high North Atlantic is considered to be one of the more "sensitive" parts of the climate system. The deep water formation areas are only a few scant hundred kilometers east of the East Greenland current, which drains low salinity water from the Arctic Ocean. Should some new climate create new ocean circulation patterns so that the whole area became flooded with low salinity water, deep water formation would cease, the surface would freeze in winter (as does the Bering Sea now, although it is at lower latitude), and Europe would lose its benign winter climate.

Land has yet another influence. If winds are such that they cause surface water to move away from the coast, it is usually replaced from below; "upwelling". The deeper water is cooler than the surface water it replaces, and it in turn cools the lower atmosphere, thus strongly influencing the local coastal climate.

The Ocean is Water

Water is one of the essential requirements for life. It is also a powerful solvent and the ocean contains, in solution, measurable quantities of all of the elements. Thus in principle all that is required for plant life, and the whole food chain dependent on it, is light. Light also is available in the surface layers except in high latitudes in winter.

In practice, however, life is self-limiting. Planktonic plants are grazed by planktonic animals which in turn provide food for larger animals. Debris from all of these organisms - fecal pellets, dead bodies, fragments - sinks. With it goes the carbon compounds and nutrients of which it is composed. Thus the surface layers become depleted in carbon compounds and in nutrients. There is so much carbon in the ocean that life is never inhibited by its lack, even when depleted. However the loss of nutrients usually severely limits plant growth.

On the way down, the sinking debris provides food for mid-water animals and is decomposed by bacteria. Excreta from these organisms adds carbon compounds and nutrients to the water, while their respiration consumes oxygen, so the deeper layers become enriched in carbon compounds and nutrients and depleted in oxygen. It was noted above that deep water is slowly rising, pushed up from below. Thus the oldest water (in the sense of time since it was last ventilated to the atmosphere) is that water lying just below the surface layers. This is the water that has been longest accumulating carbon compounds and losing oxygen, so it is the most enriched in the former and impoverished in the latter.

It is the surface layers which are exposed to the atmosphere, and which tend to come into equilibrium with the atmosphere. Thus atmospheric carbon dioxide is not related to the average concentration of carbon compounds in the ocean, but to the impoverished levels of the surface waters. Without this biological activity in the ocean, the concentration of atmospheric carbon dioxide would be substantially higher than it actually is.

In upwelling regions, deeper water comes to the surface, carrying with it its load of carbon compounds and nutrients. The nutrients enable vigorous growth to take place, which absorbs at least some of the excess carbon. The remaining carbon can be exuded to the atmosphere as carbon dioxide. It

should be noted that at least some of the water upwelling to
the surface sank, perhaps centuries before, in a bottom-water-
formation process in cold high latitudes. The saturated vapor
pressure of carbon dioxide over cold water is much higher than
that over warm water. Thus upwelled water, particularly as it
is warmed by the sun, can be expected to give off carbon
dioxide. The amount of carbon dioxide absorbed in high
latitudes by cold water, and given off in low latitudes when
the water is again warmed when it returns to the surface
layers after having completed its long stay in the deep ocean,
greatly exceeds that given off by fossil fuel burning.

The Ocean and Climate Change

The above description identifies effects of the ocean on
climate, and on the atmospheric carbon dioxide content.   What
happens when the load of greenhouse gases, including carbon
dioxide increases, and as a result the climate changes?
Unfortunately our knowledge of the details of the way in which
the system works is so incomplete that much must be
speculation. Two major experiments are under way, designed to
greatly increase that knowledge: the World Ocean Circulation
Experiment, designed to provide a huge increase in the amount
of data we have on the deep ocean, and the Joint Ocean Fluxes
Study, designed to give much more complete information on the
biological processes of importance.   Both these experiments
started within the last year, and it will be some years before
their impact on our understanding is really felt.   In the
meanwhile, let us speculate!   In speculating I shall
unashamedly use the first person singular.
     A new climate will mean new wind fields, so the wind-driven
circulations will change.   So will the associated boundary
currents and the location and intensity of upwelling. It will
also mean new precipitation patterns.   Whether these changes
could produce the "North Atlantic catastrophe" discussed above
is questionable, and there is no special reason to believe
that they should. However the possibility cannot be excluded
with existing evidence.
     The ocean may be expected to warm.   How fast? Enhanced
greenhouse-gas calculations indicate that at present we are
experiencing an increase of heat flux to the surface of about
2 watts/m compared with the preindustrial situation. That
much heat would be enough to warm the upper 500 m of the ocean
(roughly the part of the ocean in the "surface circulations")
by about one third of a degree/decade. That has certainly not
been happening. Observations do not permit us to believe in a
present rate of warming of as much as that.   Suppose the whole
ocean were warming.   There is no suggested mechanism for this
to occur; nevertheless one can calculate that the 2 watts/m
would raise the temperature by 0.04° /decade.   Again, the
evidence is against it.   Also, the expansion of the ocean
associated with an extra 2 watts/m would cause average mean
sea level to rise by about 4 mm/year, almost regardless of
where that heat were put in the ocean.   The actual rise is
agreed to be not more than 2 mm/year, and is probably less,
and not all of that is believed to be associated with warming
the ocean.

Thus I believe that the much-talked-about enormous heat capacity of the ocean is not, at present at least, very important in delaying global warming.

If the extra heat is not being stored, how can it be got rid of. Increased evaporation is a good bet. At present, average evaporation, although not very accurately known, is believed to be about 1 m/year. This requires 80 watts/m , so 2 watts/m amounts to 2.5%. No measurements of evaporation or of precipitation are sufficiently accurate either to support or to refute a suggestion that there has been an increase in both of 2.5%.

Saturated vapor pressure over water increases by 6%/°. The half degree temperature rise generally assigned to the last century therefore produces a 3% increase in saturated vapor pressure. To one with a micrometeorological background like myself, it is quite normal to expect evaporation rate to change directly according to vapor pressure. However, the large scale atmospheric modelers find a negative feedback which reduces this effect by a factor of about four, so the comforting agreement between increased surface heating and increased vapor pressure may not be all that significant. (Nevertheless, I personally rather like it!)

If it turns out that the modelers exaggerate their negative feedback on evaporation, and evaporation does closely follow vapor pressure, then the 5 watts/m additional downward flowing energy to be expected by enhanced greenhouse effect in the middle of the next century could be nicely accommodated by a 1° surface temperature rise - which is at about the lower extreme of results obtained by climate modelers.

If, on the other hand, we experience the upper warming scenario, and temperatures rise by 5°, with higher values expected at high latitudes, what would then be expected? Surely deep water formation would be reduced, even if not eliminated. Deep water formation of water appreciable warmer than at present is most unlikely. Unless somehow the salinity became much higher than at present, the water would simply not be heavy enough to get down. Nevertheless, winter cooling at high latitudes would continue to be very intense. If no major changes in salinity occur, deep water formation, of water near the freezing point as today, would be expected on a reduced scale, starting later and being less intense. Thus there might be some modified, attenuated version of the "North Atlantic catastrophe", which however would be much more supportable because of the general high latitude winter warming.

While most of this discussion has involved the northern hemisphere, where most of the people live, in fact about two thirds of the deep water formation occurs in the southern hemisphere. The processes there are also subject to modification with changing climate. There also is reason to believe that global warming would result in decrease bottom water formation.

With a continued, if reduced, supply of cold bottom water, the deep layers of the ocean would remain cold, although not quite so cold as today. The general rate of rise in the interior of the ocean would be reduced, and with it the resupply of nutrients to the surface layers. Thus some reduction in biological production might be anticipated.

What about carbon dioxide? At present, although measurements are inadequate to prove it, we can assume that the ocean is absorbing carbon dioxide by two mechanisms:

(1) The upper layers, which come into repeated contact with the atmosphere even if only every few years, more or less track the atmospheric concentrations and absorb their share of the increase in atmospheric concentration.
(2) Newly formed deep water carries down with it carbon compounds in concentrations corresponding to present day atmospheric values. The water newly pushed up into the surface layer was last at the surface hundreds of years ago, and accordingly has lower values of carbon compounds.

These two processes, neither of which is known with any accuracy, seem to be about equal in importance and may be enough to account for what is usually assigned to the ocean in the total carbon budget.

If deep water formation were much reduced, the first process would be largely unaffected, but the second would be reduced, apparently resulting in an increased atmospheric accumulation of carbon dioxide. On the other hand the reduced vertical velocity would decrease the rate at which carbon-enriched water was returned to the surface. Despite reduced biological productivity, this could result in more sequestered carbon, with less passing to the atmosphere.

All this can be modelled. However with our present weakness in both understanding and data it is not evident that modelling would produce any more convincing results than essentially verbal arguments of the kind given here.

Mankind's Greenhouse Effect: Is It Already Here?

W.W. Kellogg

National Center for Atmospheric Research (Retired),
Boulder, CO, USA

The addition to the atmosphere of infrared-absorbing (greenhouse) gases by the activities of mankind is changing the composition of the atmosphere, and theory tells us that this must cause a global warming. The worldwide burning of fossil fuels and the production of carbon dioxide is currently the single largest influence on our global temperature and climate, and other manmade trace gases are also contributing to the greenhouse effect. As a result of the increase of these greenhouse gases Planet Earth has warmed as a whole by 0.5 to 0.6 K in this century, and will continue to warm. Other environmental changes are also beginning to take place gradually, such as changes in distribution of rainfall and snowfall. All of these will have an impact on mankind. The community of nations is now debating what action, if any, it is prepared to take to slow the inevitable climate change.

1. INTRODUCTION

The theory of the greenhouse effect is one of the best understood in the field of meteorology, and it has been studied for about two centuries. Simply stated, the theory takes into account the fact that certain gases in the atmosphere absorb some of the outgoing infrared radiation from the surface, radiation that would otherwise escape to space. Fortunately for us all, the absorption of some of the infrared radiation by the atmosphere keeps the surface of the Earth about 33 K warmer than it would be without those greenhouse gases. Thus, without the greenhouse effect our planet would be perpetually frozen over, and life would probably not exist here.

While the greenhouse theory is well established, the idea that mankind is actually changing the climate of the Earth by its addition of greenhouse gases to the atmosphere is not universally accepted. Some argue that the global system that determines our climate is too complex to understand, and that our theoretical climate models run on supercomputers are simply not capable of simulating "the real thing". Others, expressing a certain amount of wishful thinking, argue that there may be factors in the climate system that will work to counteract the greenhouse warming, and they seek "negative feedback mechanisms" to prove their

case. They like to dismiss the prospect of global climate
change by maintaining that the climate system is very
"robust".

In this paper I propose to briefly review the facts in
the matter, including evidence that a global warming is
already taking place, and then explain our current
understanding of how mankind can affect the vast system that
determines the Earth's climate. Finally, I will discuss the
matter of what actions the countries of the world might be
able to take to slow the global warming if they really wanted
to. A decision to take such actions will depend on a value
judgement: When will the predicted climate change be
considered by a majority of the people to be "unacceptable"?

2. OBSERVED GLOBAL CHANGES ALREADY TAKING PLACE

The sun obviously warms the surface of the Earth, and
the Earth eliminates this heat energy by emitting infrared
radiation back to space. There is thus a long-term balance
between the incoming solar radiation and the outgoing
infrared radiation, a balance that determines our mean
temperature. Can mankind affect this balance?

The answer is clearly yes. We are changing the
concentration of certain trace gases in our atmosphere that
absorb some of the outgoing infrared radiation, and therefore
less radiant energy escapes to space. This warms the lower
atmosphere.

The most important "greenhouse gas" by far is water
vapor, but we will not talk much about it. The next most
important is carbon dioxide, and that we will talk about a
great deal. Also important are several other trace gases,
notably methane, nitrous oxide, and the infamous
chlorofluorocarbons (CFC's). As we add these greenhouse
gases to the atmosphere their atmospheric concentration
increases, since they are mostly chemically stable and have
long residence times, and the result is a turning up of the
greenhouse "thermostat" and a consequent further global
warming.

Carbon dioxide is produced when we burn fossil fuels or
wood, and some is also released in the production of cement.
Over geologic time there is a long-term balance between the
release of carbon dioxide by volcanoes and sea floor
spreading, and the removal of carbon dioxide from the
atmosphere to form carbonate rock--the latter subsequently
volatalized by the internal heat of the Earth and returned to
the atmosphere in the form of carbon dioxide [1]. Now we are
influencing that long-term balance by our enormous use of
fossil fuels and (to a lesser extent) by cutting and burning
trees in the tropics.

Here are a few facts: In 1989 we released about 6
gigatons (billions of tons) of carbon in the form of carbon

dioxide by our fossil fuel burning. Since the OPEC embargo of 1973 the rate of this release has gone up about 2% per year, a doubling time of 35 years (before that it had been going up at about 4% per year). Carbon dioxide is a chemically stable gas that remains in the atmosphere for a very long time, and its main sink is the world ocean. Over the past thirty years or more (until very recently) about 56% of the carbon dioxide produced by fossil fuel burning has remained in the atmosphere, not taking into account the additional carbon dioxide released by deforestation (whose magnitude is uncertain). The remainder has been taken up by the oceans [1][2].

The longest continuous record of carbon dioxide concentration is that obtained by C.D. Keeling on the slopes of Mauna Loa, on the Island of Hawaii. This long record, starting in 1958, shows a more or less steady increase of carbon dioxide concentration of about 0.35% per year (which has increased in the past three or four years); and if we go back before the Industrial Revolution began the total increase to the present has been almost 30%. Records of observations of carbon dioxide have also been kept at Point Barrow, American Samoa, Australia, the South Pole, and elsewhere, and they all give the same story--the concentration of carbon dioxide is steadily increasing [3].

The atmospheric concentration of methane, another greenhouse gas, has been increasing about three times faster that carbon dioxide--about 1% per year. It is a very strong absorber of infrared radiation, and its importance as a cause of global warming will equal that of carbon dioxide early in the next century if current trends persist [4]. Still another source of greenhouse warming is the CFC's, increasing at about 3% per year, but their increase may slow down if the countries of the world abide by the "Montreal Protocol" that legislates a phasing out of their production (for reasons having to do with the destruction of the ozone layer--but that is another story).

With such an increase in the greenhouse gas concentration it is not surprising that the global mean surface temperature, measured at both land stations and ships for the past 120 years or more, has increased by 0.5 to 0.6 K [5][6]. The increase has been less in the tropics and more in the Arctic; and the warming has been greater over land areas than over the oceans. The trend shows up about equally in both hemispheres, but there are some differences in the year-to-year and decade-to-decade rates of change. These shorter term changes are due to a combination of volcanic eruptions that put particles into the stratosphere, in the total output of radiation from the Sun (we now know that it is a slightly variable star), and complex changes in the interactions between the atmosphere and oceans (one well known example being the quasi-periodic El Nino events). It has also been claimed that the nuclear tests of the late 1950's and early 1960's may have put enough nitric oxide into the stratosphere to cause a temporary cooling [7]. However, it must be confessed that there is a great deal left to be explained concerning the global response to all these various external influences.

Climatologists have been searching for other pieces of evidence concerning the greenhouse warming. For example, our theoretical calculations show that there should be a cooling in the stratosphere along with the warming in the troposphere, and indeed a small stratospheric cooling has taken place in the forty years or so that upper air observations have been made at enough stations to be representative [8]. It was expected that the warming would cause polar ice in both hemispheres to retreat, but this has not occurred to any significant degree [9]. The most recent experiments with dynamic climate models shows why: during the early stages of the warming the North Atlantic and the North Pacific are now expected to actually cool for a while due to changes in the circulation pattern bringing cooler air over those northern oceans; and in the Southern Ocean the answer may lie in the changes in ocean mixing as the Earth grows warmer [10][11].

The most heralded and widely discussed changes have been, and will continue to be, changes in rainfall and soil moisture. These will be discussed in Section 3, and it will be shown that there have been dramatic changes of rainfall distribution in the past few decades.

These facts about what has been occurring in the world should be convincing enough evidence for a thoughtful person to see that a climate change is on its way, albeit in the early stages . In the next section we will discuss such observations in the light of our theoretical understanding, the purpose being to determine the degree to which the real world and the theoretical world of our climate models are consistent with each other.

3. CLIMATE THEORY, PREDICTIONS, AND LESSONS FROM THE PAST

While the essence of the greenhouse theory itself has a long history, with the greatest early contributor being Svante Arrhenius [12], more or less complete models of the entire climate system are a relatively recent development. They had to await (among other things) the availability of supercomputers to simulate the complexities of this system. At the same time, research on past climate changes have given us a vast fund of information and new insights about the way our system behaves . Perhaps most significant, scenarios of past climate changes have given us a chance to test the reliability of the models by comparisons with "the real thing".

Modern Climate System Models

With our powerful computers and rapidly improving data on the physical factors involved in such a calculation, experiments with state-of-the-art climate system models have given us a powerful tool for predicting the effects of increasing greenhouse gas concentration . The five most advanced models predict a 2 to 4 K warming for an equilibrium calculation of the response to a doubling of carbon dioxide. A dynamic calculation, in which the change in greenhouse gas concentration is introduced gradually in a more realistic way, shows a delay of two or three decades in the warming due to the thermal capacity of the oceans--it simply takes a few decades for the upper part of the oceans to warm up, and the deep oceans will take centuries to respond [10][11][13].

This is not the place for a detailed explanation of how climate models work. Indeed, they represent one of the most ambitious applications of computers, and they are exceedingly complex. In brief, they integrate the seven time-dependent equations governing the motions, conservation of mass, thermal conditions of the atmosphere, and distribution of water vapor and liquid water by taking 10 to 20 minute time steps and applying these equations at each of some 10,000 to 20,000 grid points or boxes over the globe, such grid boxes being set at around ten levels in the vertical (the number of levels varies from model to model, as does the size of the grid). Not only is the changing atmosphere analysed, but the oceans must be included as well. The shapes of the land and mountain ranges are also specified in the model. They not only create maps of winds, temperatures, clouds, precipitation, and so forth at each time step, but they also keep track of such factors as snowcover on land, sea ice distribution, sea surface temperature, and soil moisture . The atmospheric part of a climate model, known as a general circulation model, or GCM, is quite similar to the computer programs used by the weather forecasters to determine the next few days' weather.

There are a number of climate models being developed and improved, most of them in the U.S. and one in the United Kingdom [14] (others that I know about are being developed in Canada and Germany), and they give us a great deal of information about how the climate system of the Earth can be expected to respond to a change of some "boundary condition" (such as greenhouse gas concentration, solar output, changes in the Earth's orbit around the sun, and so forth). Although these theoretical models are run on the fastest computers available and take as many of the factors in the climate system into account as human ingenuity will allow, they are clearly highly oversimplified compared to "the real thing". Thus, we have good reason to wonder whether they can be relied upon to give a truly reliable and credible simulation of our climate system.

The point to keep in mind, however, is that they are the best tools we have at this time. Furthermore, we can test them in a number of ways against the real system, and this gives us both a measure of our success and suggestions for how to improve them further [2][15]. In what follows we will discuss some of these tests of model results against lessons from the past.

Shifts of Precipitation and Soil Moisture

While the temperature of a region is certainly important in determining where things can grow, it is really precipitation and soil moisture that are the most crucial. That is why farmers and ranchers and wood producers are so concerned about rainfall and evaporation, and why droughts are so devastating to agriculture of all kinds.

Our more advanced climate models not only calculate where precipitation will occur, but take the next step and calculate how much evaporation and runoff will take place at each grid point. Thus, it is possible in a climate model experiment to take the difference between the soil moisture on a model Earth with doubled carbon dioxide and the same model Earth with conditions as they are now. This reveals where it may, according to the models, become wetter or drier in the future. Some results for five such climate model experiments are presented in Figures 1 and 2 for North America and the region dominated by the Asian monsoon circulation [14][16].

We have already admitted that we are not very confident that our climate models are capable of reliably simulating the response of the Earth to a doubling of greenhouse gas concentrations, and calculations of the distribution of soil moisture are particularly difficult. Fortunately we have a way of checking on the model results, and that is to ask the paleoclimatologists to describe the conditions during periods in the past when the world was warmer than normal, either in this Century or in the more distant past.

On the whole, such comparisons have proven very gratifying. Agreement between the model warmer Earth, shown in Figures 1 and 2, and maps of conditions during warmer periods in the past do not show agreement everywhere and in all seasons of the year, but the larger scale patterns at mid-latitudes of North America and Eurasia are fairly similar in summertime. Perhaps most significant from a practical standpoint are the indications of summertime droughts in the middle of the continents, and generally more moist conditions in wintertime. Also, the monsoon circulations and their resulting summer rainfall were shown to be intensified in the model experiments, in agreement with the history of rainfall in such places as the Rajastan Desert of northwest India. We will not discuss these results in more detail here, since they have been reported elsewhere [14][16][17].

Foreseeing Future Climate Changes and Their Time Scale

We have seen that the models predict a global average warming of 2 to 4 K when the doubling of carbon dioxide takes place. This condition will be delayed by the thermal lag of the oceans, but hastened by the addition of other greenhouse gases, notably methane and the CFC's. The warming in the Arctic is expected to be two or three time greater than in

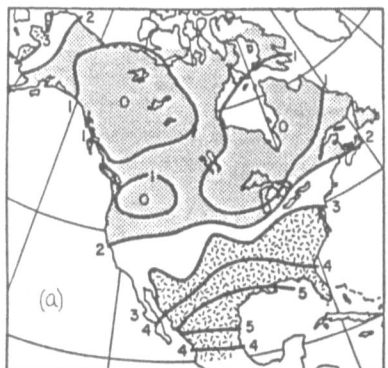 

Figure 1. Map of North America showing the degree of
agreement among the five climate models on the direction of
the soil moisture change (a) in winter and (b) in summer with
a doubling of atmospheric carbon dioxide concentration.
Areas shaded with hen-scratches show where three or more of
the models agreed on a decrease of soil moisture; areas
stippled with small dots show where three or more agreed on
an increase. (Source: Kellogg and Zhao, 1988 [14])

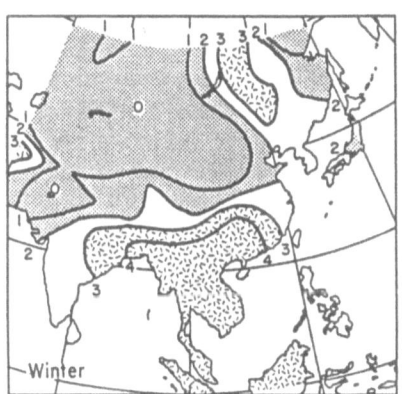 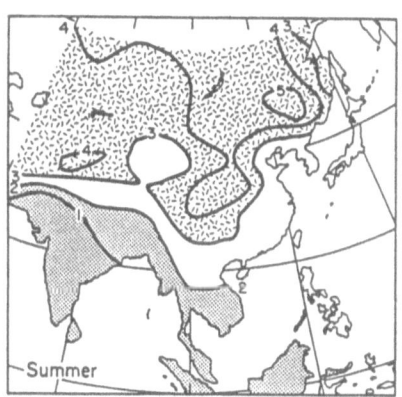

Figure 2. Same as Fig. 1, but for the regions of Asia that
are dominated by the monsoon circulation. (Source: Zhao and
Kellogg, 1988 [16])

the tropics. Along with this warming will be changes in the
global circulation patterns that determine temperature and
rainfall distributions, and there will be shifts of the
probability of droughts and floods.

We have not mentioned the rise of sea level, but, of
course, any sea level rise as a result of the warming will
have important impacts on every coastline. In spite of some
rather overplayed fears of a slippage (or surge) of the
massive West Antarctic ice sheet and an iminent rise of sea
level by 5 to 7 m (!), now the consensus of the glaciological

and oceanographic community is that in the next century we may expect about a 1 m rise (perhaps even less) [18]. It is believed that this would be due primarily to the expansion of the sea water, mostly in the upper part of the oceans, as it gradually grows warmer . In the Netherlands they are currently using this prudent estimate in their program of dike improvement and maintenance.

Returning now to temperature and precipitation, so far we have geared the discussion to what will happen when carbon dioxide atmospheric concentration is doubled and the other greenhouse gases are correspondingly increased . But when will that occur?

The answer depends to some extent on some biogeophysical factors, such as how the ocean will behave as a sink for these trace gases, how ocean plankton will respond, how the Arctic tundra may either decay or grow as it grows warmer, the probability of major volcanic eruptions, changes in solar activity, and so forth.

However, the chief player in this drama is mankind. We must ask how humanity will be using fossil fuel in the decades ahead . We must ask whether alternative energy sources will be either proven economicly competitive to fossil fuels or mandated by governments. We must ask whether nuclear power will gain favor in the U.S., Japan, Europe, and elsewhere.

All these questions are clearly in the domain of economics and politics, but I have tried to bracket the range of possibilities by invoking a "high" scenario and a "low" scenario of future fossil fuel consumption . The high one assumes a continued growth worldwide of fossil fuel use at 2% per year, which is about the same rate of growth as observed since 1973. The low one assumes that we will rapidly reduce our dependence on fossil fuels (even in the Third World), so that global consumption drops back to the present level in fifty years. I believe the first is too high, and the second is too low, but they are useful bounds to allow us to talk about the time scale of climate change.

If you believe the high scenario, then you should expect the doubling to occur a little before 2050, as shown in Figure 3. That means an expected rate of increase of global mean surface temperature of almost 1 K per decade. Following the last ice age, which climaxed about 18,000 years ago, a 1 K warming generally took place over a period of 1000 years or more--regional rates of warming may have been as much as ten times faster, however, or possibly 1 K per century. It is the possibility of such a rapid change in the future, unparalleled in the long history of mankind, that worries ecologists and agriculturalists. Natural ecosystems usually could not move fast enough to remain in their climatological niches? Would farmers and ranchers be able to adjust in time? Will the change of climate result in a loss of our forest plantations before they can be harvested? These are but a few of the problems that come to mind [19].

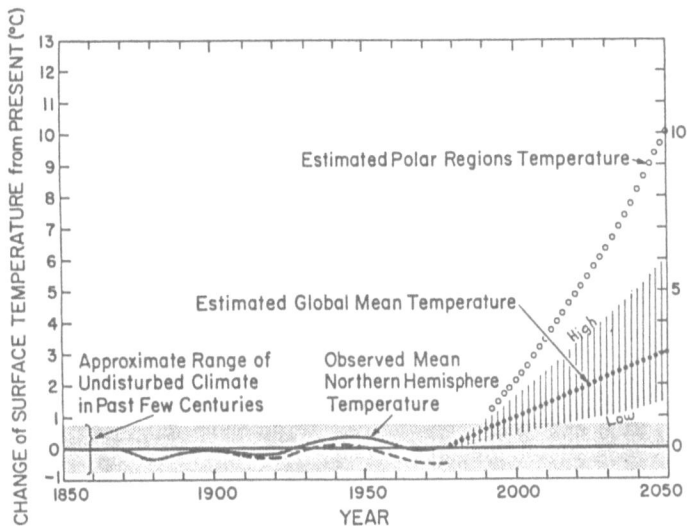

Figure 3. Past and future globally averaged temperature and
estimated polar regions temperature. The dashed line
indicates the temperature record that might have taken place
if there had not been greenhouse gases added to the
atmosphere. The vertical bars indicate the range between a
"high" and "low" scenario of fossil fuel use (see text).
This temperature scenario assumes that a doubling of the
greenhouse gases will occur about 2050 for the dotted
"estimated temperature" curve, but it could be later or
earlier depending largely on the rate of fossil fuel use.

    If you prefer the low scenario, then you can buy
another fifty or so years, and the doubling may not take
place until around 2100. It will continue to rise as long
as we go on burning fossil fuels. It seems highly unlikely
that the countries of the world will agree to cut back on
their use more rapidly than I have hypothesised in the low
scenario. So let us then look forward to a minimum rate of
warming of <u>about</u> 0.5 K <u>per</u> <u>decade</u>.

    Climate modelers are aware of some factors not taken
into account adequately in their models, as I have said.
Both the models and the real system have a large number of
feedback mechanisms that can hasten the change (if positive)
or slow the change (if negative). My impression is that the
important negative feedbacks are already included, but that
there may be some rather important positive ones that are
still left out--feedbacks that would hasten the warming when
they come into play [20].

    Here are two examples of positive feedbacks that are not
taken into account in our current climate models. They both
have to do with changes in the climate system that can affect
the fluxes of carbon dioxide, and consequently its

concentration. First, consider the way in which the oceans take up the added carbon dioxide that we put in the atmosphere. Each year less than half of the added carbon dioxide goes into solution in the upper ocean, as disolved gas, as bicarbonate, or as carbonate, and then it is slowly stirred downwards into the intermediate and deep ocean water where it can reside for centuries [1]. If this downward transport and diffusion did not take place, the upper ocean would come to a near equilibrium with the atmosphere, and little further carbon dioxide would be taken out of the atmosphere. As the upper ocean warms there will be two effects on this takeup: The increased stability caused by warm water over cold water would slow the stirring downward, and, secondly, warm water cannot absorb as much carbon dioxide as cold water. Thus, with the oceanic "sink" for atmospheric carbon dioxide slowing down, more will remain in the atmosphere--the airborne fraction will then be greater, and there is some indication that this may already be taking place [21].

The second positive feedback that I will mention has to do with the carbon dioxide and methane locked in the deep deposits of peat that cover the taiga and tundra of the Arctic. As the climate warms the summer melting season will grow longer, and more of the surface layer of peat will melt and be drained of its water. That will permit the peat to decay and release its carbon dioxide, and the methane that is now locked into the peat in the form of hydrates, or clathrates, will also be released. Thus, the warming will cause an increase in the greenhouse gas concentration, and that will in turn cause a further warming [19].

These two positive feedback mechanisms are mentioned here in order to show that there are a variety of complex interactions in the climate system that we do not adequately deal with in our current models, but they may be important [20]. We have singled out two that would act to accelerate the warming, and there may be others (not yet clearly identified) that could either accelerate or slow the warming. As our climate models continue to improve, and the computers available have more power, we will incorporate more and more of these factors into the models. That is our major challenge.

4.   WHAT CAN WE DO TO COPE WITH CLIMATE CHANGE?

Again we must emphasise that the greenhouse warming of the Earth is due to mankind, in particular the burning of fossil fuels, and to a lesser extent it is due to the deforestation taking place in the tropical rain forests. Contributions to the warming are also being made by our activities that produce methane, and in that case the blame probably rests with our agricultural practices such as rice growing, more herds of cattle, and encouraging termites by forest cutting in the tropics--with some further contribution due to leakage from natural gas wells and pipelines. Still

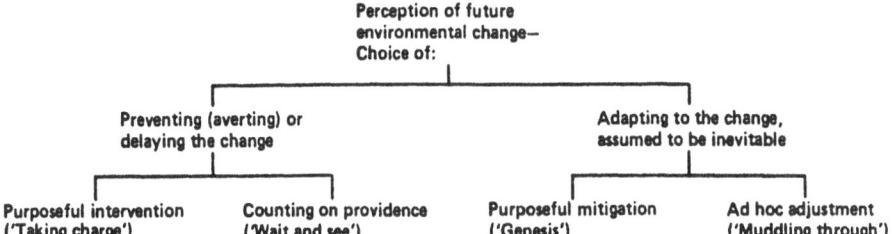

Figure 4. A decision tree showing the range of choices that could in principle be made in order to cope with climate change. The actions on the left would presumably have to be taken on a worldwide basis to be effective (see text), but those on the right can be taken at any level of society.

another source of a greenhouse gas is the CFC's that are used as refrigerants, aerosol spray propellants, cleaning electronic parts, and so forth [4].

In Figure 4 is shown a kind of decision tree displaying the range of choices open to governments, industry, agriculture, and the public as we contemplate the future of the Earth and its climate. These policy choices range from taking charge of the situation and stopping the climate change (or at least slowing it down) on the left to just muddling through, shown on the right--that's what we usually do when decisions are hard to make. And in between these two extremes is a very broad area of actions that we can take to lessen the undesirable impacts of the change, assuming that we perceive it as inevitable.

The "taking charge" option, if adopted, would mean that the nations of the world will have to reduce their consumption of fossil fuels. There have been a number of high level meetings to discuss such a policy, two recent ones being held in the Dutch city of Noordvijk in November 1989, and the next one being in Washington in April 1990. At both of these meetings there seemed to be a general agreement on the seriousness of the global environmental crisis, and the proposal (backed by most western European countries) that seemed to gain the most favor was an initiative calling for the industrial nations to freeze their carbon dioxide emissions at 1988 levels by the year 2005. While there seemed to be general support for this idea, it did not become a part of the Noordvijk Conference's final declaration, nor was it agreed to in Washington. Four economic superpowers, the U.S., the U.S.S.R., the United Kingdom, and Japan, all refused to agree to it. If China and India had been present it is very doubtful whether they would have agreed to it either.

These discussion about what to do in the face of the impending environmental crisis and global warming will continue at the Second World Climate Conference (Geneva,

October-November 1990) and at the United Nations (November 1990). In 1992 the United Nations will sponsor the Second International Conference on the Environment in Brazil.

It is not hard to fathom some of the reasons for reluctance on the part of a government to promise that it will limit the use of fossil fuels. In the case of the U.S., the U.S.S.R., and China there are enormous reserves of coal and other fossil fuels. In most industrial countries the corporations that control the production, distribution, and use of fossil fuel energy are among the most powerful, and they are not likely to welcome any government constraints. Furthermore, the less developed countries are struggling to become more industrialized, and they can hardly do this without resorting to the cheapest and most convenient energy source, fossil fuels [22][23].

On the other hand, the Noordvijk proposal and its successors does not seem to constitute a very drastic cutback on carbon dioxide production, at least for the industrialised countries. The U.S. is presently one of the most wasteful users of fossil fuel energy in the world, vying with the former East Germany for that dubious distinction with a conspicuous per capita consumption of about 5 tons of carbon per year (1986 figures). West Germany and Japan, for example, use less than half that much per capita, but have equally high living standards [24]. It seems that these countries are just a great deal more efficient in their fossil fuel use, or have resorted to alternative sources of energy.

The most obvious path to follow in an effort to cut back on carbon dioxide emissions is to conserve energy. The industrial countries are indeed making progress in this respect, and most of them have leveled off in the rate of increase of use of fossil fuels (except, in the case of the U.S., in fuels for automobiles and electric generation) [25]. There are a variety of suggestions for encouraging further reductions, such as higher gasoline taxes (the U.S. currently has about the lowest of any country), penalties for driving large gas-guzzling cars, a charge to a factory or utility for its carbon emissions, a program of introducing more efficient lighting, heating, and refrigerating equipment, better insulation of buildings, and so forth.

The biggest step forward in this respect will be taken when alternative, non-fossil, energy sources become economicly competitive. Solar, biomass (a form of solar energy in a sense), and wind are all gaining in acceptance. It is undoubtedly time to have another look at nuclear energy too, since there are some new developments in inherently safe reactors that may make them more acceptable to the public [26]. Furthermore, if hydrogen replaces gasoline and natural gas, then the release of carbon dioxide will be greatly reduced.

The important thing to note is that most of the measures to cope with a global warming and climate change make sense anyway, even if there were no impending enviromental crisis. They include, as we have said, conservation of energy, use of alternative and renewable energy sources, better water management to deal with droughts, protection of the forest resources of the world (especially in the tropics), and <u>above all</u> measures to reduce the runaway increase in the world's population.

NOTE: This article is a revision of an earlier paper by this author, published in the Proceedings of the 8th World Hydrogen Energy Conference, Honolulu, July 1990.

REFERENCES

1. Post, W.M., T.-H. Peng, W.R. Emanuel, A.W. King, V.H. Dale, and D.L. DeAngelis, 1990: The global carbon cycle. <u>American</u> <u>Scientist</u>, 78, 310-326.

2. Kellogg, W.W., 1987: Mankind's impact on climate: The evolution of an awareness. <u>Climatic Change</u>, 10, 113-136.

3. Keeling, C.D., A.F. Carter, and W.G. Mook, 1984: Seasonal, latitudinal, and secular variations in the abundance and isotope ratios of atmospheric CO2. <u>J. Geophys. Res.</u>, 89, 4615-4628; also C.D. Keeling, personal communication.

4. Ramanathan, V., H.B. Singh, R.J. Cicerone, and J.T. Kiehl, 1985: Trace gas trends and their potential role in climate change. <u>J. Geophys. Res.</u>, 90, 5547-5566.

5. Jones, P.D., T.M. Wigley, and P.B. Wright, 1986: Global temperature variation between 1861 and 1984. <u>Nature</u>, 322, 430-434.

6. Jones, P.D., and T.M.L. Wigley, 1990: Global warming trends. <u>Scientific American</u>, 263, 84-91.

7. Kondratyev, K.Ya., 1988: <u>Climate Shocks: Natural and Anthropogenic</u>. Wiley Interscience, New York, 295 pp.

8. Ramanathan, V., 1988: The greenhouse theory of climate change: A test by an inadvertent global experiment. <u>Science</u>, 240, 293-299.

9. Parkinson, C.L., and D.J. Cavalieri, 1989: Arctic sea ice 1973-1987: Seasonal, regional, and interannual variability. <u>J. Geophys. Res.</u>, 94(C). 14,499-14,523.

10. Washington, W.M., and G.A. Meehl, 1989: Climate sensitivity due to increased CO2: Experiments with a coupled atmosphere and ocean general circulation model. <u>Climate Dynamics</u>, 4, 1-38.

11. Manabe, S., K. Bryan, and M.J. Spelman, 1990: Transient response of a global ocean-atmosphere model to a doubling of atmospheric carbon dioxide. J. Phys. Oceanog., 20, 722-749.

12. Arrhenius, S., 1896: On the influence of carbonic acid in the air upon the temperature of the ground. Philosophical Magazine, 41, 237-271; see also S. Arrhenius, 1906: Worlds in the Making. Harper and Brothers, New York and London, 250 pp.

13. Hansen, J., I. Fung, A. Lacis, D. Rind, S. Lebedeff, R. Ruedy, and G. Russell, 1988: Global climate changes as forecast by Goddard Institute for Space Studies three-dimensional model. J. Geophys. Res., 93, 9341-9364.

14. Kellogg, W.W., and Z.-c. Zhao, 1988: Sensitivity of soil moisture to doubling of carbon dioxide in climate model experiments. Part I, North America. J. Climate, 1, 348-366.

15. Kellogg, W.W., 1990: Theory of climate: Transition from academic challenge to global imperative. Chapt in Greenhouse Glasnost: The Crisis of Global Warming, ed. by T.J. Minger, Inst. for Resource Management and Ecco Press, New York, pp. 93-115.

16. Zhao, Z.-c., and W.W. Kellogg, 1988: Sensitivity of soil moisture to doubling of carbon dioxide in climate model experiments. Part II, The Asian monsoon region. J. Climate, 1, 367-378.

17. Kellogg, W.W., 1982: Precipitation trends on a warmer Earth. Chapt. in Interpretation of Climate and Photochemical Models, Ozone and Temperature Measurements, ed. by R.A. Reck and J.R. Hummel, American Inst. of Physics, New York, p. 35-46.

18. NRC, 1985: Glaciers, Ice Sheets, and Sea Level: Effect of a CO2-Induced Climatic Change, Polar Research Board, National Research Council, Washington, D.C., 330 pp.; see also Villach/Bellagio Workshops, 1987: Developing Policies for Responding to Climatic Changes. Report of workshops held in Villach, Austria (23 Sept.-2 Oct., 1987) and Bellagio, Italy (9-13 November, 1987), rept. prepared by Jill Jaeger, World Meteorological Organization Rept. WCIP-1, WMO/TD-No. 225, Geneva, Switzerland.

19. Houghton, R.A., and G.M. Woodwell, 1989: Global climatic change. Scientific American, 260(4), 36-44.

20. Kellogg, W.W., 1983: Feedback mechanisms in the climate system affecting future levels of carbon dioxide. J. Geophys. Res., 88, 1263-1269.

21. Keeling, C.D., 1990: Personal communication.

22. Kellogg, W.W., and R. Schware, 1981: <u>Climate Change and Society: Consequences of Increasing Atmospheric Carbon Dioxide</u>, Westview Press, Boulder, Colorado, 178 pp.; see also W. W. Kellogg and R. Schware, 1982: Society, science, and climate change. <u>Foreign Affairs</u>, 60, 1076-1109.

23. Schneider, S.H., 1989: <u>Global Warming: Are We Entering the Greenhouse Century?</u> Sierra Club Books, San Francisco, 275 pp.

24. Marland, G., 1989: <u>Fossil fuels CO2 emissions: Three countries account for 50% in 1986.</u> Carbon Dioxide Information Analysis Center (CDIAC), Oak Ridge National Laboratory, Oak Ridge, Tenn., p. 1-3.

25. MacKenzie, J.J., 1988: <u>Breathing Easier: Taking Action on Climate Change, Air Pollution, and Energy Insecurity.</u> World Resources Inst., Washington, D.C., 23 pp.

26. Haefele, W., 1990: Energy from nuclear power. <u>Scientific American</u>, 263(3), 136-145.

Strategies for Environmentally Sound Economic Development: A Framework for Interdisciplinary Collaboration

F. Duchin

Institute for Economic Analysis, New York University,
269 Mercer Street, New York, NY 10003, USA

Summary. Bold and massive initiatives are needed to take on the problems of poverty in the developing countries and pollution on a global scale. Economists can contribute a concrete, operational analytic framework for assessing some of the implications of alternative scenarios about the world economy. Preliminary results of a feasibility study focused on energy use and air pollution are reported in this paper: projections are presented of CO_2, SO_2, and NO_2 emissions in all regions of the world associated with combustion of fossil fuels.

Collaboration is required with engineers to incorporate descriptions of technological means of achieving economic and environmental objectives into a world economic database and with applied scientists to represent feedback between human activities and changes in the natural world such as those affecting temperatures and precipitation. The construction of the economic model and database, and these interdisciplinary objectives, need to be incorporated into the scientific research agenda.

1. Introduction

Pressures on the global environment, which are already very great, can be expected to increase substantially for many decades with even modest economic and demographic growth although many of the particulars, including the effects of environmental pollution, degradation, and transformation on human activities, are still poorly understood. Individual actions and national policies aimed at alleviating pressures on the global environment and at the same time raising the standard of living in the poorest countries are bound to proliferate and can contribute to these objectives. However, in view of the very modest success of development efforts over the last several decades and the scale of potential environmental modification, it seems clear that only bold and systematic initiatives can substantially change what would be the outcome of another century of business as usual. These initiatives need to be operational at four different levels. They require a workable institutional framework and adequate financing. I do not

The author is Director of the Institute for Economic Analysis at New York University and Research Professor at the Robert F. Wagner Graduate School of Public Service. The work described here is partly funded by the United Nations. The views expressed are those of the author, and no endorsement on the part of the United Nations is implied.

The Global Environment
Editors: K. Takeuchi · M. Yoshino © Springer-Verlag Berlin, Heidelberg 1991

attempt to address these two requirements in this paper but offer as an example the Marshall Plan, rather successful for both donor and recipient nations, which was innovative in the concreteness of its objectives, worked through new and existing, interlocking institutions, and allocated over its short life more funds than did the World Bank and IMF together in that period.

Initiatives for resolving these problems need to be operational at two other levels that are more squarely in the domain of the scientific and technical community: they need to be both feasible and economic. The physical feasibility of a specific objective requires that there are one or more ways in which it can be achieved. Targets like 25% reduction in carbon dioxide emissions or sustained average annual rates of growth of 3% or more in real per capita consumption in developing countries are common points of reference. However, even if such targets are parts of binding agreements, compliance is difficult to enforce unless specific, practical implementations can be reasonably demonstrated. Each different, feasible way to proceed will have a host of specific consequences including economic ones. To the extent that the major consequences of alternative potential courses of action can be anticipated, before they are launched, this evaluation provides a basis for obtaining political and financial support that will be necessary for their implementation.

The desirability of some potential initiatives may appear to be self-evident; this is the case for many instances of energy conservation through improvements in efficiency which can rely mainly on market mechanisms after an initial transition period. Other initiatives may require significant coordination, new research results, major financing, and long lead times, and are bound to have unanticipated side-effects. This might be the case for massive forestation projects in the tropics, for example. Realistic feasibility studies are indispensable, especially in the latter but also in the former case.

Such a study requires an analytic framework in which, first, the present situation can be systematically described and, second, alternative courses of future action can be realistically evaluated. A general research plan has been outlined by the author (Duchin, 1990a) and can now be made more concrete on the basis of recent work.

2. Strategies and Scenarios

Economic and environmental objectives are often at odds. There is clearly an international mandate to pursue both, but with more urgency attached to the environment in the rich countries and to development in the poor countries. A global perspective can help in determining priorities for the international community and in perceiving common problems and common solutions. Some critical areas related to development and atmospheric pollution -- the raw material for national and world strategies -- are identified in this section.

Improving the productivity and expanding the agricultural resource base requires massive investment in order to reverse degradation, increase and sustain yields of food and fuelwood, and possibly contribute to sequestering of atmospheric carbon, especially in the tropics. The state of the resource base and its potential will be specific to each region, but increased use of fertilizers and irrigation are likely to be common requirements.

Any development strategy will require massive quantities of capital goods. Modern, efficient combustion equipment of a standard capacity and small, standardized computers will be the main components of such a package as these are the building blocks for modernization and expansion of industrial capacity (and the provision of electricity in particular). Motorized transport equipment for carrying goods and passengers will also be required in potentially vast quantities in all developing regions. Key questions that need to be answered are how this equipment will be absorbed into each economy, where it will be produced, and who will pay for it.

Benefits from conservation and recycling can be important especially in the rich countries because of the scale of use of materials. After a transition period in which new markets are established, these practices should meet little resistance because they tend to reduce costs. Attempts to rationalize transportation, on the other hand, meet considerable resistance, especially proposals for reduced reliance on the personal car. Vastly improved fuel efficiency and possible fuel alternatives can resolve some of these problems over the medium term, but increased reliance on practical and convenient common transport systems also has an increasingly important role to play in both rich and poor countries.

Increased investment in prospection for fossil fuels in developing countries could dramatically alter the geographic distribution of known reserves. This would alleviate economic problems in these countries and surely intensify their use of energy with attendant environmental effects. Increased processing of fuels prior to combustion, for example conversion of fossil fuels (or biomass) to methanol, could have both environmental and economic advantages. Widespread use of methanol, or other non-traditional fuels, would require a transitional period for the establishment of markets and the conversion of related physical infrastructure and would have far-reaching international economic implications. Greatly increased reliance on processed biomass for energy appears to be realistic for tropical countries.

Let us assume the existence and generous funding of a World Program for Environment and Development (WPED). A group of countries applying to WPED for assistance would need to present a concrete strategy for approval. This is similar in spirit but considerably more complex than a project proposal to, say, the World Bank or (in an earlier period) the submission of production and trade targets to the European Recovery Program (Marshall Plan). Very few countries have experience even in formulating, let alone actually implementing, multi-faceted strategies. The Japanese experience, however, is exceptional and Japan could perhaps provide leadership in this area.

An analytic framework such as the model and database described below is intended to play the crucial role of evaluating possible strategies. A strategy would need to be converted to an operational scenario by identifying the sectors of the economy involved and quantifying related inputs and outputs. Each scenario can be analyzed for consistency and feasibility and evaluated in terms of its various economic and environmental implications.

3. Model of the World Economy

Such an analysis requires a conceptual framework situating human activities within the natural world. The fundamental contribution of the economist is the ability to model the world economic system as a whole while maintaining significant detail about the functioning of its individual parts. Economists have a long history of evaluating the feasibility of concrete individual projects, usually in monetary costs and benefits. We have a shorter history of analyzing broader scenarios, and to date these have generally been extremely simple and vaguely specified. It is possible at this time to build a model of the world economy that can be useful for the purposes under discussion, but this will require a major research effort far more ambitious than its predecessors. The world economy needs to be disaggregated into separate but interacting regional economies each of which is itself represented in terms of the full range of economic activities taking place in that region. An appropriate level of disaggregation strikes a balance among several objectives: the realism provided by additional detail, the challenge of developing corresponding data, and the need to maintain a fully operational framework as the work proceeds.

From a theoretical point of view there are three fundamental requirements of a working model of the world economy that have not been adequately satisfied by past versions. The first is for separate but integrated analyses of physical stocks and flows on the one hand and of costs and prices on the other. Only in this way can the economist lay the groundwork for absorbing the information provided by engineers and applied scientists whose concerns will generally be directly represented in physical terms. The associated costs, where appropriate, are evaluated in price computations. Economists' duality theories provide for physical/price systems, but in empirical work it has become standard practice to combine them in a single system expressed only in monetary values.

The second requirement is that the model be truly dynamic: this involves not only intersectoral consistency but also intertemporal consistency, especially with regard to physical capital. The model needs to be able to trace a realistic time path rather than being constrained by one of the two most common idealizations: balanced growth (where all sectors grow at the same rate) or movement from one static equilibrium to another. Economists have experience with dynamic models which now needs to be incorporated into an operational model of the world economy.

The distinguishing characteristic of a world model lies in the treatment of international trade and capital flows. Great strides in the theory of international exchange are so far barely reflected in empirical investigations, especially those that go beyond descriptions of past trade flows. In a model like the one I have described, with distinct physical and price models for each region, it should finally be possible to determine both international volumes of trade and international prices thus providing a firm theoretical basis for the representation of trade.

4. World Economic Database

As the "information age," made possible by computer-based automation, proceeds, many scientific communities are organizing to create, integrate, and standardize the often massive databases which are fundamental for their disciplines. The structure of a model of the world economy provides a conceptual organizing device for the kind of database that will be needed to study the economics of sustainable development. Problems, however, abound. Because the collection of economic data is highly decentralized and measurement units and techniques poorly defined, volumes and quality of data collected at different times in different countries are highly variable. In addition, a great deal of information that is required for the effective use of a world economic model is simply not available. While the challenge of building such a database is considerable, so is the motivation as it will serve many other purposes besides supporting a model of the world economy.

The subject of standardizing international socioeconomic databases for integration into a library of scientific databases bearing on global climate change has begun to be raised within the broader scientific community. However, even the major existing economic databases do not satisfy the quality control standards familiar to physical scientists. World economic databases rely upon country-reported data which are not independently verified and that differ in definitions, conventions, classifications, accounting methods, sources of data, level of detail, completeness, documentation, and so on.

Despite their shortcomings, significant amounts of economic data have been compiled from these official (although unverified and relatively unstandardized) country reports by the United Nations and a few other international organizations, and these are an indispensable starting point for a world economic database. There are also unofficial economic databases, both proprietary and ones that are made public. For example, major industries sometimes organize "trade associations" that compile and publish data about international production, imports, and exports of the goods and services of interest to them (motor vehicles, aircraft, machine tools, etc.) Such data have the distinct advantage that they are generally measured in physical units of relatively disaggregated goods and services. These are an important supplement to the official data and need to be included, selectively, in an emerging world database.

There is a great deal of economic information about the past that is not systematically collected or estimated: the use of firewood for fuel is one example of information that should be included in an emerging database, along with other energy raw materials, even if only rough estimates can be made.

Appropriate and realistic quality control criteria for economic data need to be worked out. Closer adherence to contemporary international standards for National Accounting data could naturally be anticipated if more resources were available for these purposes especially in developing countries. This would improve the scope, accuracy, timeliness, and to some extent the international comparability of the data. However, it will always be problematic to directly confirm individual figures. And it will always be problematic to integrate or even compare data from different countries especially when they are expressed only in value terms (because of differences in currencies[1]).

Economic data, in value and in physical units, need to be documented by the primary compiler and screened for plausibility by the compiler of international data. The standards, documentation, and screening characteristics are the basis for judging data quality. However, errors -- even significant ones -- will always creep in for a host of well-known reasons that it is not necessary to repeat here. If sources of economic data were disqualified when individual figures were found to be off by more than any given percentage, it would simply be impossible to construct a useful database. We have to accept the fact that these data are estimates, some very rough.

These observations raise the tactical question of how to allocate resources to data collection. My feeling is that it is necessary to first design the structure and then complete an initial draft of the database even if many figures are crude estimates. Subsequently, those figures most critical for particular purposes need to be refined, and the outcomes of analyses using the database can pinpoint the parts of the database most in need of further refinement. The success of this approach depends upon a realistic conceptual model and a relatively stable mathematical implementation, a reasonable although approximate database, and an international and interdisciplinary community of researchers committed to the program. These are not easy to furnish, but there is no better alternative.

When it comes to the future, data have been collected in numerous specialized studies but there are no comprehensive and systematic projections of economic data. Individual research teams need to be encouraged to standardize and document the projections of key variables and parameters developed in the course of their work for inclusion in the world database. However, preparation of the data required for the systematic analysis of the kinds of scenarios described earlier will require a considerable level of effort. This effort is likely to be forthcoming only it appears that nations might, in fact, embark on that course of action.

Developing such scenarios requires a political, social, economic, and technical understanding that can be furnished only by local analysts. Their participation is indispensable: they are the only source for the empirical content of the scenarios. If their priorities are reflected in the scenarios, they will be motivated to develop the data required to describe and analyze them. These are also bound to be the most realistic and promising scenarios.

The idea of carrying out periodic large-scale surveys of experts to identify strategic technical objectives is most advanced in Japan where it has worked extremely well in the economic sphere. An international survey of this type could be carried out using the

---

[1]Conversions based on the exchange rate, while a common practice, cannot be defended in a scientific database. The single exchange rate between two currencies generally governs the ratio of prices in the two countries of traded items but is not directly related to prices of untraded goods and services.

electronic network conferences on sustainable development that are quickly being established. In addition to the information it generates, this process also helps create a constituency for the subsequent implementation of specific scenarios.

5. A Feasibility Study: Preliminary Results

Since June 1989, researchers at the Institute for Economic Analysis have been developing the design and a preliminary implementation of a World Model and World Database to analyze the environmental and economic implications of alternative scenarios for the UN Conference on Environment and Development in Brazil in May 1992. This work is intended to demonstrate the feasibility and the importance of a more ambitious study.

In this framework, the world is divided into 15 geographic regions and each is represented by about 50 economic sectors. The model is of the input-output type which means that each sector of the economy is described in terms of the detailed inputs produced by each of the 50 sectors, as well as detailed capital and labor inputs, required to make a unit of its characteristic output. For the past, this input structure reflects the mix of technologies already in place. For the future, input structures are projected according to alternative technological assumptions.

Virtually all models of the world economy and even of individual national economies which are applied in empirical analysis include an input-output module as it is the most compact way to introduce a consistent representation of individual sectors of the economy. In most of these models, the input-output detail is used to disaggregate the economy-wide stocks and flows (like investment or consumption) while the latter are determined only at the aggregate level. The input-output framework offers the possibility of building these stocks and flows from the bottom up on the basis of the kinds of decisions taken more or less simultaneously in individual sectors of the economy. The potential of this framework has not yet been fully exploited because doing so requires extensions to existing models and especially to existing databases. In clearing the way for this new work, we have attempted to explicitly address the most restrictive limitations of simple input-output models.

Over the past several decades, virtually all applied input-output studies have been carried out using the basic, static input-output model. This model has many unique advantages which account for its continued use. Its major limitations are its static nature (a characteristic shared by many economic models) and the absence of feedback from prices to the structure of production. Much of the work of the Institute for Economic Analysis during the last decade has been aimed at extending the static input-output framework to an operational dynamic model and to developing the interaction between the physical and price models, both conceptually and in applied work.

While dynamic input-output models are familiar in the theoretical literature since the late 1940's, a fully operational version that assures intertemporal as well as intersectoral consistency in the production and use of capital goods (Duchin and Szyld, 1985) and its implementation (Leontief and Duchin, 1986) are more recent. This model takes an economy off the knife's edge path of balanced growth models by allowing for unused capacity in individual sectors which are in this way freed to grow at different rates. The model has now also been implemented in several European countries. In work that has just been completed, a compatible dynamic price model and an entirely new dynamic input-output income model have been formulated and applied (Duchin and Lange, 1990).

Only in the simplest applied exercises have input-output coefficients been literally "fixed," that is, assumed to be the same in different years, different countries, or different scenarios. However, the practice of allowing individual coefficients to change automatically (i.e., endogenously) in response to changes in relative prices has been resisted by input-output economists out of concern that the resulting vectors of sectoral inputs will generally be infeasible from the point of view of actual production since, in reality, inputs cannot be substituted one by one.

In most input-output studies, whether comparative static or dynamic, input coefficients (for a specific country, year, and scenario) are provided exogenously and do not change in the course of a computation. In recent work a middle ground has been sought between the two, equally unrealistic extremes of complete, automatic substitution and no automatic substitution (Duchin and Lange, 1990). This is accomplished by taking a larger unit for price-based substitution than individual inputs: sets of inputs comprising distinct, technologically feasible alternatives. We have identified specific situations in which we are prepared to incorporate automatic, price-based choice of technology in our future work, mainly in the manufacturing sectors; the situation is far more complicated in agriculture, mining, and in many service sectors. Progress will be made in this important but difficult area by incorporating within the input-output framework the price-based choice among alternative input structures, especially in the case of new facilities.

Another point of departure is the input-output model of the world economy built by Wassily Leontief and his colleagues in the 1970's (Leontief, Carter, and Petri, 1977). We have begun incorporating the advances of the last decade into this World Model (Duchin and Lange, 1989a and b; Duchin, Lange, and Johnsen, 1990a, b and c). The basic mathematical structure of an initial version of the new model (see Annex A) is being implemented in stages while maintaining an operational model at each stage.

One of the fundamental strengths of the input-output frame-work, and of the closely related area of activity analysis, is the potential for the direct representation of distinct technologies and organizational arrangements in the vectors of detailed inputs to each sector. However, despite several decades of history, this potential has been barely tapped because of the unanticipated difficulty of translating expertise about production into not merely a symbolic representation but systematically defined, quantified, and documented databases of input coefficients. A major objective of the current work is to demonstrate the feasibility of this task and to make headway in carrying it out.

The principal source of input-output data has historically been the official government tables which are extracted from censuses and surveys based on national accounting principles. Data work over the past decade at the Institute for Economic Analysis and in other places, however, has shifted increasingly from accounting toward technical sources. The technical literature and working engineers can provide the basis for quantitative descriptions of future technological alternatives. Examples are provided in Leontief and Duchin (1986) and Duchin (1990b).

The first stage of the feasibility study involves a Baseline Scenario covering the period from 1980 to 2000. Most of the work to date has gone into updating key parameters and variables and extending the database to include sector- and region-specific parameters for each time period governing emissions of carbon dioxide and oxides of sulfur and nitrogen associated with the combustion of coal, gas, and petroleum products. Given energy use estimates by fuel computed by the model, carbon dioxide coefficients are related directly to the energy content of each fuel. Sulfur oxide emission coefficients reflect the sulfur content as well as the energy content of the fuels in use, the mix of refined petroleum products, and any pre- and post-combustion removal of sulfur. The nitrogen oxide emission coefficients reflect the mix of fuels and the characteristics of the boilers (especially the firing mode) in which they are burned. Each of these parameters has been assembled or estimated for all regions for 1980 from a host of sources and projected for 1990 and 2000. (See Duchin, Lange, and Johnsen (1990c) and Lange (1990) for the figures, sources and related discussion.)

The Baseline Scenario computations produce estimates of emissions by region in 1980 and projections for 1990 and 2000 (see Annex B) which are consistent with projected changes in regional levels of population and economic activity and in economic structure. The scenario presently incorporates population and economic projections made by the United Nations.

The computed carbon emissions match the estimates made by other researchers very closely. In the case of sulfur and nitrogen oxides, however, ours are the first global estimates: these are compared with other sources in Tables 1 and 2. The comparisons need to be made with caution for many reasons; the main discrepancy is that the other sources do not include all countries in a region.

Our sulfur oxide emissions match those of other sources reasonably closely for the regions for which comparisons exist (Table 1). The principal observed discrepancies are explained by the facts that the other sources include for the US 3.5 million tons of SO_2 due to industrial processes other than combustion; post-combustion controls in Japan are underestimated in our work; for some regions, like Low-Income Asia and Major Oil Producers, only one or two countries are included in the other sources. The greatest discrepancy, however, cannot be explained away. Taking account of the heavy reliance on the low quality and high sulfur content of coal in Eastern Europe, our estimate is twice that of the other sources. Assuming a 1.5% sulfur content of coal, which is surely a lower limit, our estimate is still more than 50% higher. The sulfur content of coal in Eastern Europe and the Soviet Union is discussed in Duchin, Lange, and Johnsen (1990c).

Other estimates of nitrogen emissions are even spottier than for sulfur. While the same caveats hold, the comparisons with our results are also relatively good where other estimates exist.

Over the two decades from 1980 to 2000, annual worldwide emissions of carbon (in the form of carbon dioxide) due to the combustion of fossil fuels can be expected to increase by nearly two-thirds from 4.7 to 7.8 billion tons according to the Baseline

Table 1. Baseline Scenario Estimates of Anthropogenic Sulfur Oxide
Emissions by Region in 1980 Compared to Other Sources
(10^6 tons of SO_2)

Region	UNEP/WHO	UN/ECE	Combined	World Model Baseline Scenario
1 High-Income North America	27.81[a]	27.85[a]		23.97
2 Newly-Industrializing Latin America	na	na		2.18
3 Other Latin America	na	na		0.35
4 High-Income Western Europe	18.24	13.52	18.44	15.00
5 Medium-Income Western Europe	4.46	3.92	5.45	5.15
6 Eastern Europe	12.57	8.83	14.07	29.47 (22.31[b])
7 USSR	na	na		28.00
8 Centrally-Planned Asia	14.21	na		13.16
9 Japan	1.64	1.26		2.09[c]
10 Low-Income Asia	2.23	na		4.14
11 Major Oil Producers	0.20	na		0.95
12 Other Middle East and North Africa	0.29	na		0.23
13 Sub-Saharan Africa	na	na		0.13
14 Southern Africa	na	na		2.14
15 Oceania	1.48	na		1.47
World Total	na	na		128.43 (121.27[b])

na   not available

[a]Includes 3.5 million tons due to industrial processes other than combustion.
[b]Assumes a fuel sulfur content of coal of 1.5%, rather than 2.0%, in Eastern Europe.
[c]Post-combustion pollution controls are underestimated.

Notes:

1. These comparisons need to be made with caution. The UNEP/WHO and UN/ECE estimates do not cover all countries in each region, and the coverage of activities resulting in reported emissions is not uniform. Some countries include industrial emissions from sources other than combustion, but in most such cases (except for the US) the magnitudes involved are small.

2. Three regions are covered by both sources but with significant geographic gaps. In the column called "combined" we have aggregated the emissions for countries reported in only one of the sources and the higher of the two estimates, so as to include more economic activities, for countries included in both.

Sources: UN Environment Program (UNEP) and World Health Organization (WHO), 1988, p.908; UN Statistical Commission and the Economic Commission for Europe (ECE), 1987, Table I-15

Table 2. World Model Estimates of Anthropogenic Nitrogen Oxide Emissions by Region in 1980 Compared to Other Sources (10^6 tons of NO_2)

Region	UNEP/WHO	UN/ECE	Combined	World Model Baseline Scenario
1 High-Income North America	22.21[a]	22.03[a]		20.07
2 Newly-Industrializing Latin America	na	na		2.90
3 Other Latin America	na	na		0.66
4 High-Income Western Europe	10.73	8.51	10.80	11.90
5 Medium-Income Western Europe	1.40	1.09	1.53	2.35
6 Eastern Europe	na	2.96		3.90
7 USSR	na	na		3.57
8 Centrally-Planned Asia	4.40	na		5.85
9 Japan	1.34	1.26		3.16[b]
10 Low-Income Asia	na	na		3.89
11 Major Oil Producers	0.10	na		1.40
12 Other Middle East and North Africa	0.11	na		0.47
13 Sub-Saharan Africa	na	na		0.21
14 Southern Africa	na	na		0.61
15 Oceania	0.92	na		0.89
World Total	na	na		61.83

na   not available
a   Includes emissions due to industrial processes other than combustion.
b   Post-combustion pollution controls are underestimated.

Notes:

1.   These comparisons need to be made with caution. The UNEP/WHO and UN/ECE estimates do not cover all countries in each region, the coverage of activities resulting in reported emissions is not uniform. Some countries, notably the US, include industrial emissions from sources other than combustion, but in most cases the magnitudes involved are small.

2.   Three regions are covered by both sources but with significant geographic gaps. In the column called "combined" we have aggregated the emissions for countries reported in only one of the sources and the higher of the two estimates, so as to include more economic activities, for countries included in both.

Sources: UN Environment Program (UNEP) and World Health Organization (WHO), 1988, p.908; UN Statistical Commission and the Economic Commission for Europe (ECE), 1987, Table I-15.

Scenario (see Table 3), emissions of sulfur oxides (measured as SO_2) by 70% from 128 to 219 million tons, and emissions of nitrogen oxides (measured as NO_2) by over 50% from 65 to 100 million tons. Over this period emissions of all pollutants grow most rapidly in the developing regions while the shares generated in the rich, developed regions systematically decline. This last result reflects underlying assumptions about the developing regions: continued high population growth and higher than average economic growth especially in India, China, and the oil-rich Middle East, accompanied by the rapid incorporation of production and consumption practices that are more energy-intensive than traditional alternatives, e.g., electric equipment and appliances, cement and other construction materials, and automobiles.

Global emissions of the three pollutants increased much more slowly than overall world economic activity measured as the sum of regional GDP's) between 1980 and 1990: annual carbon emissions grew 15% over the decade, sulfur oxides 19% and nitrogen oxides 11% (see Table 3). This reflects the effects of energy conservation practices in the 1980's in the rich developed regions which offset the enormous increases in energy use in the fastest growing developing economies.

Baseline Scenario projections for the decade from 1990 to 2000 show a rate of increase of global pollutant emissions equivalent to the rate of growth of global economic activity: annual carbon emissions grow by 44% over the decade, sulfur oxides by 48%, and nitrogen oxides by 40% (while world GDP increases 41% or at an average annual rate of 3.5% real growth). While improvements in energy efficiency and pollution controls used in the 1980's were assumed under the Baseline Scenario, no additional improvements have yet been incorporated into the Baseline Scenario for the period after 1990.

Coal releases considerably more carbon per unit of energy content than petroleum or gas and is used on a large scale in some of the largest economies: the US, USSR, China, and India. These facts account for the larger than average increase in carbon emissions in 3 of the 4 regions in which these countries are included. (See Table B.1 in Annex B.) Carbon emissions associated with fuel use in the developing countries are currently underestimated by the Baseline Scenario (but with no effect on sulfur oxides emissions and very little on nitrogen oxides) because we have not yet included biomass fuels, which emit more carbon per unit of energy content than the fossil fuels.

Because of a heavy reliance on coal with its high sulfur content relative to oil and gas, in particular the low quality, high sulfur coal of some countries and the lack of effective pollution controls, Eastern Europe and the Soviet Union account for an enormous share of global emissions of sulfur oxides. According to the Baseline Scenario they generate 45% of the world's sulfur oxide emissions in 1980 and 40% in 2000, compared to only about 25% of the world's carbon emissions (see Table 3).

In the rich economies, roughly half the nitrogen oxide emissions result from the use of motor vehicles whereas nitrogen oxide emissions in developing regions are generated mainly by combustion in industrial settings and the production of thermal electricity. The use of petroleum in motorized transport results in an order of magnitude of emissions of nitrogen oxides per unit of energy content greater than in most other uses.

Table 3. Shares of Pollutant Emissions by Regional Groupings from the Combustion of Fuels Under the Baseline Scenario in 1980, 1990, and 2000

Regional Group	Carbon			SO_2			NO_2		
	1980	1990	2000	1980	1990	2000	1980	1990	2000
Rich, Developed Economies	50.2%	42.9%	38.0%	32.0%	28.1%	23.9%	54.4%	48.3%	43.8%
Developing Economies	20.1	26.0	31.2	16.5	23.4	28.8	17.7	23.8	28.5
Eastern Europe and USSR	24.7	26.0	25.2	44.8	40.8	40.0	22.0	22.3	22.2
Other	5.0	5.1	5.0	6.8	7.7	7.3	5.9	5.7	5.5
World Total (%)	100.0	100.0	100.0	100.0	100.0	100.0	100.0	100.0	100.0
World Total (weight)[a]	4.7	5.4	7.8	128.4	148.6	219.2	64.6	71.61	100.0

[a] 10^9 metric tons of carbon, 10^6 metric tons of SO_2, 10^6 metric tons of NO_2

Note: Rich, Developed Economies include World Model regions 1, 4, and 9. Developing Economies include regions 2, 3, 8, and 10-13. Eastern Europe and USSR include 6 and 7. Other includes 5, 14, and 15. Regions 1 through 15 are identified in Tables 1 and 2.

Source: Tables B.1-B.3 in Annex B.

Projections of future nitrogen oxide emissions are particularly sensitive to future transportation practices, especially growth in the use of motor vehicles in developing countries and motor vehicle pollution controls, the latter mainly in developed countries. While a moderate increase in motor vehicle use in developing countries is anticipated under the Baseline Scenario, a dramatic increase is possible and would result in comparably greater increases in nitrogen emissions. At the same time, additional controls over the emissions of nitrogen oxides in the rich developed countries are highly likely and would reduce their share of global emissions even further. The latter observation also holds for sulfur emissions.

6. Interdisciplinary Collaboration

The next task is to extend the time horizon of the Baseline Scenario for several decades beyond 2000 by representing plausible business-as-usual assumptions about structural changes over this period with attention to development and energy use in particular. In parallel, "experimental" scenarios need to be constructed with some preliminary results in time for the Brazil meeting in May 1992. In order to transform potential strategies for dealing with poverty and environmental degradation into operational scenarios, engineering and other technical expertise is needed to describe and quantify the inputs and outputs associated with alternative technological approaches. A functional level of detail will be somewhat more abstract than an engineering manual and considerably more disaggregated than an economist's production function. The conceptual framework will need to be reconciled with the empirical content in the course of this collaboration.

Models of the world economy, such as the World Model, and climate models (today's General Circulation Models or GCM's) still have challenging agendas to accomplish in their separate domains. I do not believe that the time is yet ripe for conceptual and operational integration. But it is not too early for systematic interaction with accompanying establishment of common concepts and vocabulary. Of course, this early period will surely involve efforts to "link" the two types of models in at least a loosely defined association.

The basic difficulties for this collaboration are as fundamental as different time horizons and time steps and different geographic units and methods of spatial coverage.

While these divergences will eventually need to be reconciled, it may be more fruitful to start by identifying the areas of overlap. For the GCM's, emissions of atmospheric gases associated with human activities (still essentially limited to CO_2) are exogenous and presented as an average annual rate of global increase. A World Model can provide spatially disaggregated estimates for the past and for the future for all gases of interest corresponding to alternative scenarios. To complete the feedback loop, the World Database includes assumptions about the productivity of agriculture, fisheries, and forestry which will clearly be influenced by regional changes in temperatures, precipitation, and so on, which are the outcomes of a GCM scenario. Yet another community of applied scientists is developing the process models that relate these climate changes to biological productivity, including crop growth, which directly affect economic productivity.   All these pieces will eventually need to be coordinated to help us begin to imagine, and effectively deal with, the likely long-term consequences of our actions.

References

Duchin, F. 1990a. "Evaluating Strategies for Environmentally Sound Economic Development: An Input-Output Approach." Innovation and Input-Output Technique, Journal of the Pan-Pacific Association of Input-Output Studies, Japan.

Duchin, F. 1990b. "Framework for the Evaluation of Scenarios for the Conversion of Biological Materials ands Wastes to Useful Products" Structural Change and Economic Dynamics, Vol.1, No.2.

Duchin, F., with G. Lange. 1990. "Technological Choices and Their Implications for the U.S. Economy, 1963-2000: Report on the Construction and Application of an Engineering/Input-Output Model and Database."   Draft final report to the National Science Foundation for grant #ENG - 8703347. March.

Duchin, F. and G. Lange. 1989a, 1989b. "Strategies for Environmentally Sound Development: An Input-Output Analysis." Progress reports #1 (October) and #2 (December) to the United Nations, UN Contract #PTS/CON/66/89.

Duchin, F., G. Lange, and T. Johnsen.   1990a, 1990b, 1990c. "Strategies for Environmentally Sound Development..." Progress reports #3 (March),#4 (June), and #5 (September) to the UN.

Duchin, F. and D. Szyld. 1985. "A Dynamic Input-Output Model with Assured Positive Output." Metroeconomica, Vol. 37, pp. 269-282.

Lange, G. 1990.   "The Role of Technology in Representing the Emission of Atmospheric Pollutants for a Model of the World Economy," working paper.

Leontief, W., A.P. Carter and P. Petri.   1977. The Future of the World Economy, New York: Oxford University Press.

Leontief, W. and F. Duchin. 1986. The Future Impact of Automation of Workers, New York: Oxford University Press.

United Nations Statistical Commission and the Economic Commission for Europe. 1987. Environmental Statistics in Europe and North America. New York: United Nations.

United Nations Environment Program and World Health Organization. 1988.   Assessment of Urban Air Quality. New York: United Nations.

Annex A. Mathematical Model

This annex provides the algebra for the world dynamic physical model. The one-region dynamic price model, the one-region dynamic income model, and the one-region, cost-based choice of technology model have also been developed. All these models and their empirical application to the US economy are described in detail in Duchin and Lange (1990) and references therein. The model shown below represents technological change but does not include mechanisms for cost-based technological choices.

These modeling concepts will be integrated into the World Model according to the priorities of the specific projects funding the modeling work. The presently operational version is similar to the world dynamic physical model, but with a less detailed representation of investment.

Trade flows are computed on the basis of sectoral import coefficients (measuring the volume of imports relative to domestic output) and world market export shares. In future work, trade flows will be based on computations of comparative costs.

Definition of Variables and Parameters for Each Region

$c(t)$ capacity at beginning of period t

$c^*(t+\tau)$ capacity planned in period t for t+τ where τ is the maximum number of periods lag between delivery of a capital good and its use in production

$o(t+\tau)$ increase in capacity, if any, planned in period t for t+τ

R matrix of requirements for replacement of existing capacity

δ_i maximum admissible annual rate of expansion of capacity for sector i

$\hat{m}$ diagonal matrix of import coefficients

e vector of exports

A, x, y defined as usual (A is the matrix of current account requirements per unit of output, x is the vector of outputs, and y is the vector of deliveries (other than imports and exports) to final users.

$B^\theta(t)$ matrix of those capital requirements per unit increase in output that need to be produced in period t for first use in period $t + \theta$.

Initial Conditions

$c(t_0)$

$c^*(t), \quad t = t_0+1, \ldots, t_0+\tau-1$

$x(t), \quad t = t_0-\tau, \ldots, t_0-1$

157

The initial conditions are completed by computing o and then c, according to the equations below, for $t = t_0+1,\ldots,t_0+\tau-1$.

Model

For $t = t_0,\ldots,t_T$

$$c^*(t+\tau) = \min\left[1+\delta_i, \frac{x_i(t-1)+x_i(t-2)}{x_i(t-2)+x_i(t-3)}\right]^{\tau+1} \cdot x_i(t-1)$$

$$o(t+\tau) = \max[0, c^*(t+\tau) - c(t+\tau-1)]$$

$$c(t+\tau) = c(t+\tau-1) + o(t+\tau)$$

$$[I-A(t) - R + \hat{m}(t)]\,x(t) = \sum_{\theta=1}^{\tau} B^\theta(t)\,o(t+\theta) + e(t) + y(t).$$

Closure for World Trade

The preceding equations, one for each region, are supplemented by the following, preliminary closure for world trade (dropping the time subscripts):

$$\sum_{i=1}^{r} \hat{m}_i x_i = \sum_{i=1}^{r} e_i = \sum_{i=1}^{r} \hat{S}_i u$$

where i indexes the r regions
$\hat{S}$ is a diagonal matrix of world market export shares
u is the vector of the levels of total international trade (endogenous).

Table B.1 Carbon Dioxide Emissions by Region from the Combustion of Fuels Under the Baseline Scenario in 1980, 1990, and 2000 (10^9 metric tons)

Region	Carbon (10^9 metric tons)			Growth (1980=1.00)	
	1980	1990	2000	1990	2000
1 High-Income North America	1.39	1.40	1.80	1.01	1.30
2 Newly-Industrializing Latin America	0.15	0.18	0.26	1.20	1.30
3 Other Latin America	0.06	0.05	0.08	0.83	1.33
4 High-Income Western Europe	0.75	0.71	0.86	0.95	1.15
5 Medium-Income Western Europe	0.12	0.12	0.17	1.00	1.42
6 Eastern Europe	0.36	0.38	0.55	1.06	1.53
7 USSR	0.81	1.03	1.42	1.61	1.75
8 Centrally-Planned Asia	0.40	0.63	4.24	1.58	3.10
9 Japan	0.23	0.23	0.31	1.00	1.35
10 Low-Income Asia	0.21	0.36	0.55	1.71	2.62
11 Major Oil Producers	0.08	0.14	0.27	1.75	3.38
12 Other Middle East and North Africa	0.03	0.05	0.07	1.67	2.33
13 Sub-Saharan Africa	0.01	0.01	0.02	1.00	2.00
14 Southern Africa	0.06	0.08	0.11	1.33	1.83
15 Oceania	0.06	0.08	0.11	1.33	1.83
World Total	4.73	5.43	7.81	1.15	1.65

Source: World Model computations (preliminary).

Table B.2 Growth of Sulfur Oxide Emissions by Region from the Combustion of Fuels Under the Baseline Scenario in 1980, 1990, and 2000

Region	SO_2 (10^6 metric tons)			Growth (1980=1.00)	
	1980	1990	2000	1990	2000
1 High-Income North America	23.97	27.26	35.10	1.14	1.46
2 Newly-Industrializing Latin America	2.18	2.62	3.89	1.20	1.30
3 Other Latin America	0.35	0.26	0.46	0.74	1.31
4 High-Income Western Europe	15.00	12.85	15.14	0.86	1.01
5 Medium-Income Western Europe	5.15	6.81	9.28	1.32	1.80
6 Eastern Europe	29.47	31.94	47.39	1.08	1.48
7 USSR	28.00	28.67	40.17	1.02	1.44
8 Centrally-Planned Asia	13.16	21.50	42.38	1.63	3.22
9 Japan	2.09	1.61	2.23	0.77	1.07
10 Low-Income Asia	4.14	7.90	12.55	1.91	3.03
11 Major Oil Producers	0.95	1.23	2.80	1.30	2.95
12 Other Middle East and North Africa	0.23	0.55	0.91	2.39	3.96
13 Sub-Saharan Africa	0.13	0.09	0.16	0.69	1.23
14 Southern Africa	2.14	2.87	4.21	1.34	1.97
15 Oceania	1.47	1.82	2.52	1.24	1.71
World Total	128.43	148.61	219.19	1.16	1.71

Source: World Model computations (preliminary).

Table B.3 Growth of Nitrogen Oxide Emissions by Region from the Combustion of Fuels Under the Baseline Scenario in 1980, 1990, and 2000

	Region	NO$_2$ (10^6 metric tons)			Growth (1980=1.0)		
		1980	1990	2000	1980	1990	2000
1	High-Income North America	20.07	19.91	25.47		0.99	1.27
2	Newly-Industrializing Latin America	2.34	2.90	3.83		1.24	1.64
3	Other Latin America	0.79	0.66	1.04		0.84	1.32
4	High-Income Western Europe	11.91	11.40	13.85		0.96	1.16
5	Medium-Income Western Europe	2.34	2.18	2.81		0.93	1.20
6	Eastern Europe	3.92	4.20	6.06		1.07	1.55
7	USSR	10.27	11.75	16.17		1.14	1.58
8	Centrally-Planned Asia	3.57	5.85	11.57		1.64	3.24
9	Japan	3.16	3.27	4.51		1.04	1.43
10	Low-Income Asia	2.71	3.89	6.00		1.44	2.21
11	Major Oil Producers	1.40	2.92	4.87		2.09	3.48
12	Other Middle East and North Africa	0.47	0.63	0.99		1.34	2.11
13	Sub-Saharan Africa	0.21	0.20	0.34		0.95	1.62
14	Southern Africa	0.61	0.79	1.14		1.30	1.87
15	Oceania	0.89	1.11	1.53		1.25	1.72
	World Total	64.64	71.63	100.00		1.11	1.55

Source: World Model computations (preliminary).

Conjectures About the Chances
of Ultimate Global Sustainability

H. Krupp

Department of General Systems Studies, The University of Tokyo,
3-8-1 Komaba, Meguro, Tokyo 153, Japan

Our technological, economic and political thinking is dominated by the growth paradigm. However, societal learning at the local, national and global level may eventually lead to greater acceptance of the sustainability paradigm.

In the industrialized countries, within time spans of about 50 to 70 years, sufficient energy saving may technologically and economically permit exclusive reliance on renewable energy sources. It will take longer in developing countries.

By contrast, environmental destruction resulting from numerous polluting emissions, with annual outputs between megatons and kilograms per country, and from many projects of infrastructure development an other investments, is very difficult to control. Therefore it seems safe to assume that environmental deterioration will continue for at least another century; all the more so, if developing countries continue to grow in population and income.

Ultimate sustainability, in a probably much more artificial world than today, is imaginable only if societal learning leads to a profound self-restraint of the economy and of the profit drive, and if a high rank is attributed to the social and (environmental) costs of projects.

Introduction

Beyond global warming, this paper addresses the wider question of whether it is possible to reduce pollution to a level compatible with global sustainability.

The term sustainability, however, is vague and may be defined anywhere between two extremes:
- Adaption to a rather undisturbed Nature similar to that existing in the pre-industrial age;
- The synthesis of a homeostatic technological world where Nature would be an ensemble of biocultures for the provision of food, of biomass for energy generation, and of chemical products (from mass polymers to perfumes and drugs), together with industrial and residential areas and artificial parks, possibly with a few old "natural" parks left in low population density areas. In this case, cultural norms would have changed from present standards of "undisturbed naturalness" to norms which accept nature as a human artefact which contains less bio-diversity and reduced aesthetic complexity.

Any definition of sustainability must take into account the fact that 10 billion people, or more, would have to be comfortably provided with material resources and energy for very long periods of time.

The Global Environment
Editors: K. Takeuchi · M. Yoshino © Springer-Verlag Berlin, Heidelberg 1991

The following deliberations are more compatible with the second definition than with the first.

This paper discusses the prospects of societal learning to cope with energy provision and environmental protection, and the pertinent time constants involved.

Dichotomic Discourse

In past ages technology served the interests of feudal rulers, backed by state religions. After the industrial revolution technology became one of the major instruments of capital increase. For the last two centuries this has created growing wealth for people in the industrialized countries. However today, the profit-driven wasteful large-scale deployment of technological products and processes, unmitigated by wider issues of welfare, is destroying our natural resources at alarming rates. As a result of growing environmental awareness, our present-day techno-economic related political thinking in the highly industrialized countries is dividing along two paradigms, that of growth and that of global sustainability.

These two philosophies are complementary, but not mutually exclusive, as some technologies may be put to multiple use so that new developments in products and processes which further sustainability may simultaneously constitute growing business and sources of income. In terms of economics, the zone of overlap is characterized by investments in energy saving, renewable energy sources, and new low-emission production processes. These need to be processes which reach break-even point within periods acceptable to private households and business, either with or without social costs being taken into account.

The growth paradigm is signalled by terms such as "innovation", "technological progress", "individual entrepreneurship", "profits" and "marketing". Orientation towards global sustainability is indicated by such terms as "protection of natural resources", "emission control", and "recycling of materials". Indicators such as private and public research budgets, capital investments, content analysis of the media and private consumption patterns[1] show that the growth philosophy is at least 5 to 10 times more widespread than that of sustainability.

The growth discourse occurs in a societal feedback system of government, business and consumers, providing self-referencing frames of mind and interdependencies which condition the decision-making of society. The result is that decision makers may experience both a fatal course of societal evolution and subjective freedom of choice at the same time. An entrepreneur with choices between alternative options of capital investment, technology, and marketing at his disposal may, on the other hand, feel competition forcing him to maximize profits to the exclusion of other targets (such as minimizing social costs). Politicians are conditioned by their power game among small volatile majorities, strong lobbies, deficient global solidarity, and narrow time horizons compared with those involved in matters of sustainability. Individually, they are not accountable for the final impacts of their incremental decisions.

However, our world is not static, but is subject to
societal learning, so that in principle this philosophic split
could be overcome.

Time Contrasts of Societal Learning

The signals within the publicized information noise are being
modulated by short-term fads and fashions, but on the other
hand, major paradigms may take a long time to be generated, to
change and to die.

It seems appropriate to measure the time constants in
matters of the utilization and protection of natural
resources, especially of energy and the environment, in units
of one human generation, that is to say 25 years. Let me call
this time unit one G. The plausibility of this proposal
derives from three sets of factors:
-    The time lag between societal perception and related
action. Examples are elements of societal learning such as
the emergence of citizen movements and green parties, the
establishment of systems to monitor environmental damage, the
change of smoking habits, and the passing of a "Clean Air
Act";
-    The phase differences of societal learning between
different countries, for example between Eastern Europe and
the Netherlands or Germany; between the South and the North,
for example Italy and Scandinavia; and between developing and
highly industrialized countries. There are wide gaps in their
resource awareness and related political action at any given
point of time;
-    The time constants of changes in the state of ecosystems
which can be as large as 100 G (CO_2 storage in the oceans) or
more (ice ages). However, particularly relevant to this paper
are the time constants of the after-effects and of ecological
restoration which occur in the diffusion of surface pollution
into the groundwater or in reforestation, for example, which
are the order of 1 G.

In discussions of techno-economic and political reactions
to environmental threats, this unit of 1 G must be kept in
mind as the minimum time span required for significant steps
of societal learning to occur.

If we add 1 G each for
-    The time lag between societal perception and the start of
related action by pioneers;
-    International diffusion;
-    The development of remedies;
-    Their deployment and diffusion,
we obtain one century as the order of magnitude of time which
is probably required for the kind of major societal learning
which might lead to world-wide systems of renewable energy
provision or to ecologically more acceptable processes of
manufacturing. This has two implications: quicker remedies
cannot reasonably be expected; political action cannot occur
too early to reduce ecological destruction. If, for
simplicity, we assume environmental destruction to proceed at
the rate of economic growth[111], it will multiply within each
generation, at least in the industrialized and the so-called
threshold countries.

In order to be more concrete, let us discuss energy provision and emissions into the environment in greater detail.

Energy

From reports of two German parliamentary commissions, respectively concerned with "Technology Assessment" and "The Global Atmosphere" it appears possible, both technically and economically, to construct a sustainable system of renewable energy provision in Germany within 60 years, i.e. about 2 to 3 G. It would be based mainly on major increases of energy efficiency together with 50% reliance upon both domestic and imported solar energy. The domestic solar electricity would be provided by photo-voltaics installed on about 20% of the roofs of German houses. The imports, in the form of electricity or hydrogen, might come all the way from the Southern Soviet Union to the Sahara and Spain, to cover a wide time zone and solar peak. Fluctuating solar and wind energy would have a share of less than 25% of electricity production, with the result that grid stabilization would be possible.

There are, however, a number of factors which would constitute major political barriers to the establishment of a sustainable energy system before the market price of fossil fuels reached well above US$40 per barrel over long time periods. These factors include lack of global solidarity; the marginal majorities of western governments; the strength of those industries, which want to continue to reap profits from sunk investments and from their present structures; the retarded awareness of more energy-efficient technology[11].

The systemic task to be achieved consists of
-    Establishing internationally concerted intervention policies for the marketing of fossil fuels, in order to manage a steadily increasing price and to stretch the reserves;
-    Organizing long term investments in energy conversion technology, grid stabilization, energy transmission and distribution, energy recycling, and co-generation;
-    Complementing the present highly decentralized energy structure with a decentralized one with appropriate energy management on all levels: individual homes, office buildings, districts, cities - yet most of them to be integrated with the power grid.

Nuclear energy, which is not sustainable, is found not to be required in this German scenario.

If well managed over sufficiently long periods of time, the transition could be smooth and without loss in GNP, in spite of a shift of some private wealth to the infrastructural sector. A large number of technological and social innovations would result.

When attempting to extrapolate this promising scenario to a global scale, three major problems arise:

(1) If equity of global energy consumption per capita at German standards is postulated, energy consumption will have to increase by a factor of about four.

(2) Probably, world population will at least double within the time span considered so that altogether, eightfold energy consumption will ensue.

(3) Shortage of investment capital in less industrialized countries and problems (such as knowledge barriers and time lags) of transfer of advanced technologies will for a few decades reduce their chances of increasing their energy supply either in part or in whole by means of sustainable technologies.

If, optimistically, we assume that the industrialized countries will achieve a sustainable energy path within, say 2 to 3 G, it will still take a few decades more to achieve equitable provision of sustainable energy world wide.

On the other hand, in time spans of the order of 100 years, the build-up of a sustainable energy regime seems probable because of the exhaustion of fossil reserves and the probable adaptability of societies to changed resource conditions. Nuclear energy, if not abandoned, will probably not contribute major portions of total energy in many countries.

Environmental Protection

Environmental degradation arises generally from two major sources:
- The destruction of natural resources (soil, water, air, fauna, flora, habitats) by construction work (buildings, roads, airfields, dams, etc.) and other types of industrialization or development (agriculture, forest management);
- The introduction of foreign substances (in type and/or quantity) into the ecosphere through a great variety of emissions from industrial and commercial activities.

Atmospheric emissions may be classified by threefold orders of magnitude[2]:
- Mass pollutants such as CO_2, SO_2, NO_x, particles, solid waste, fertilizers, and detergents are put into the ecosphere at rates of megatons per year, per industrialized country;
- Filtrates, pharmaceuticals[III], biocides (pesticides, fungicides, weed-killers, etc.), other organic materials, liquid waste are released in kilotons per year;
- Heavy metals, asbestos, halogenated hydrocarbons are released in tons per year;
- Highly toxic substances such as dioxins and furanes are introduced in amounts of kilograms per year.

Toxicities are inversely related to the emission rates.

In industrialized countries, only the emission of SO_2 has been noticeably reduced. Some others may now become stabilized (fertilizers, detergents, heavy metals, for example), but others keep on increasing such as CO_2 and NO_x. In developing countries, all types of emissions are increasing.

The increasing deployment of nuclear power plants and their fuel cycle as well as the increasing releases of genetically manipulated organisms[IV], which constitute first steps towards a commercialization of all fauna and flora on the earth, constitute major additional, non-chemical hazards.

The early successes of policies of environmental protection in some countries (Japan in particular) have been achieved by targeting mass pollutants (SO_2, NO_x, particles) emitted from major point sources which are rather easy to monitor.

By contrast, a great variety of widespread diffuse emissions is much more difficult to control. It has been estimated[3] that annually about 60,000 new chemical substances are manufactured. It is not possible to investigate their ecological effects beforehand, or those of the millions of different chemical substances now available, principally because of limited laboratory capacity, the long duration of pertinent tests, problems of simulation of econiches and their interactive complexities, together with the high costs involved.

Yet, although in principle, environmental protection from many diffuse emissions seems feasible both technically and economically, it is perhaps not so politically. By far the greatest number of pollutants have not yet even become a public issue, a first condition of political action. More fundamentally and as already said, most democratic governments have to rely on small parliamentary majorities, support from industry and from volatile public opinion. All these factors militate against a sweeping clean up.

Still, there is growing awareness and societal learning aimed at attempting to cope with these problems.

Mechanisms of Societal Learning

The global ecosystem is under stress to such an extent that human welfare is suffering both in spite of and because of economic growth. Whatever the problems of measuring welfare may be, there are estimates which suggest that in the industrialized countries the rate of welfare decline may predominate over GNP growth.

This degeneration of our earth is increasingly perceived on all levels, local, regional and international, so that environmental problems are becoming public and publicized issues world wide. This started with scientific analyses and has been assisted by the formation of a great number and variety of societal movements whose networking has led to the formation of political parties, as is the case in Germany. This in turn induced a voter swing which has changed the rhetoric of the main parties significantly so that all of them now advocate environmental protection. These movements have also networked internationally. Evidence is provided by the spectacular world-wide projects of Greenpeace and the negotiations of UNEP and IPCC.

These processes have been spurred by the formation of a satellite-connected "information society" and the focus of the media on controversial subjects.

This adaptive societal learning is based on various psycho-social factors, such as:
- Individual discontent and resulting action;
- Conflict between generations;
- Feed-back to ingroup/outgroup constellations;
- Network formation;
- Competition between social groupings (e.g. churches).

Profit-driven competition is a powerful mechanism of societal learning so that market niches opened by these movements are quickly exploited (e.g. biodegradable plastics). At the same time, opportunities for the marketing of dubious products are being generated, presenting a difficult regulatory problem.

There are many barriers to the transfer of individual and group societal learning into professional and political pathways. Here are some examples:
- Those who are among the first to discover environmental harm and unethical practices and who might check harmful innovations in research and business are in the same boat as their employers. Therefore a conflict between bread-winning and individual ethics ensues;
- Career-oriented politicians are often well-advised not to form spearheads for societal developments if they want to be safe from being shot in the back by their own kind;
- Other than political reforms, mankind has another means of adaptation to a polluted or destroyed environment, that of lowering expectations and norms - in particular because of the large time constants involved. We see this at work in very many places all over the world.

However, if public awareness of environmental destruction increases and becomes translated into political pressure, the present resource-oriented attempts of governments to re-regulate the framework in which the economy and its technology is operating may proceed.

Conclusions

(1) A characteristic paradox of the present split between philosophies of growth versus sustainability is the amount and intensity of advertising and marketing of technology, and supply push. This widens existing markets and opens up new ones and keeps them going by a sophisticated sequence of fashions, remodellings, and innovations exploiting human instincts and desires beyond those that need to be satisfied for a comfortable life style. Two centuries of societal conditioning have created an intense feedback between the profit motive of manufacturers and commerce, their exploitation of human characteristics and creation of wants, the transformation of these into demand, made possible by the remuneration of highly productive life-long work in an upward spiral which has generated our economic growth. In the industrialized societies, we are entering a wasteful throw-away regime which is not sustainable on earth.

(2) Economists claim that it would suffice to put price tags on natural resources. However, there are obvious limitations to this concept, illustrated by the following examples:
- It will be difficult to establish rational norms for price tags for the many toxic pollutants which harm the resources air, water, and soil;
- Emissions from many diffuse sources cannot be monitored sufficiently;
- Because of their harmfulness, the production and use of many substances has to be stopped (e.g. asbestos, many biocides);
- Reliance on price tags increases social inequities; an example is the popularity of high-powered heavy-weight cars which bestow prestige on the owners at considerable social cost.

- It is improbable that politicians and bureaucrats will be capable of enacting and effectively monitoring and enforcing the gigantic regulatory framework which will be required. Eventually, because of the system dynamics, a race between new ways of resource exploitation and regulatory action may occur leaving the regulators behind by a large time lag, as is the case with pharmaceuticals, for instance.

Thus, the re-regulatory task is a very difficult one.

(3) Much more relevant are the problems of the developing countries for whom growth is the last hope of survival for many millions of people. By comparison, protection of the environment takes second place. The learning processes required are much more difficult to perform and to diffuse, and the profit drive of the industrialized countries and their selection of technology transfer into the developing countries as well as debt for nature approaches do not suffice, or may even be counterproductive.

On the other hand, the example given above of renewable energy provisions for Germany shows that joint projects between the industrialized and the developing countries may provide mutual benefits. In the particular case quoted, the German energy imports would be based on annual investments of up to 40 billion DM for a decade or more with the concomitant build up of technological capabilities and income in the recipient countries.

Societal learning in the North and in the South make it increasingly unlikely that the present discrepancies of income will be maintained. The development of a common energy base and common environmental protection may provide mechanisms for their reduction.

(4) There are two alternatives: people and governments may rely on a quasi-biological adaptation of societies to a changed environment by passive societal learning and a fatalistic (or cynical) attitude of laissez-faire. The disadvantage would be more environmental destruction than in the opposite case of pro-active policies. Yet, since policy making by our governments requires pressure from the electorate, and since the environment is not a really pressing issue for most people in our industrialized countries, and because of the conflicting goals of environmental protection versus development in the developing countries, political reactions world wide to the present situation will probably continue to be weak and slow.

Therefore, it is likely that on global scale environmental pollution and destruction will continue for many decades. Possible successes of the industrialized countries will be counteracted by the growing wealth and environmental destruction of the industrializing countries.

I thank Dr Darryl Macer, Tsukuba, for stimulation and help in the preparation of this paper.

References and Footnotes

[I] Today, the annual expenses for alcohol, or tobacco, or advertizing are the same order of magnitude as those spent for environmental protection.

[II] In the highly industrialized countries, this is no longer the case with a few mass pollutants, for example SO' and

fertilizers which are discussed below. Otherwise and on a global scale, the above assumption may long remain valid.
[III] Much of the consumption of potentially harmful pharmaceuticals is probably the result of seductive marketing and is medically not helpful; it is a burden to the public health system and the production of such substances adds to the pollution of the environment. Excess use of antibiotics in humans and farm animals has led to antibiotic resistance which can result in more virulent pathogenic bacteria.
[IV] 250 had been released by mid-1990[4].

[1]   Nitsh, L., et al., Conditions and Strategies for the Construction of a Solar Hydrogen Economy, Commission of Inquiry on Technology Assessment of the German Parliament, Stuttgart 1990
[2]   Amory B. Lovins: see contribution to this volume
[3]   The Council of Experts on Environmental Problems, Expertise on the Environment, 1987 W. Kohlhammer Stuttgart/Mainz,
ISBN 3-17-003364-6
[4]   Macer, Darryl R.J. Shaping Genes, Christchurch New Zealand: Eubios Ethics Institute 1990

The World of Perceptions Versus the World of Data: Notes Towards Safe-Failing the Energy Equation

B.L.B. Wiman

Department of Environmental and Energy Systems Studies,
Lund University, Institute of Technology,
Gerdagatan 13, S-223 62 Lund, Sweden

Combinations of efficient use of energy (EUE) and renewable energy sources (RES) can contribute to finding solutions (S) to problems of large-scale environmental change. Therefore, developing technological existence proofs of solutions to the "energy equation", EUE+RES →S, is important. However, for solutions to be viable a wide range of non-technical issues has to be addressed, such as perceptions of the *modus operandi* of natural systems, risk-philosophy aspects, cultural acceptability of large-scale ecological engineering, and obstacles in the process of linking science with policy. In particular, awareness and management of the above facets and tensions will grow increasingly important as interest in harnessing qualities of natural systems expands. This paper briefly explores and exemplifies some of these aspects, in certain cases drawing on challenges and opportunities facing environmental and energy systems in Sweden.

Introduction – perspectives on constraints and opportunities

Although energy supply is not distributed on an equity basis around the globe, it is not any global-level shortage of energy supply in itself that is the basic problem. In contrast to the concerns of the 1970s, the issue now is that inadvertant and potentially hazardous environmental effects (acidification, changing chemical and physical climate, threatened biotopes and species) of energy use constrain the ways in which energy can be used. Therefore, rapidly changing perceptions in environmental science are now challenging energy policies to respond.

There is growing consensus (Fig. 1) that certain global environmental futures are not only undesirable[1], but also associated with increasing risks for discontinuous and threshold behaviour in biogeochemical and biogeophysical mechanisms of the biosphere[2]. In this context it should be noted that the crucial questions must relate to meaningful time-scales, of the order of some 100 years.[3] In addition, an important shift is taking place as to what are the actual constraints on the linkages between energy and development. Perspectives are now switching from physical-resources depletion oriented and supply-driven issues of the past decades towards the insight that the dynamics and functioning of the biospheric mechanisms constitute the actual resources endangered by depletion.

For instance, although energy supply is by no means distributed on an equity basis around the globe, it is not any global-level shortage of energy supply in itself that is the basic problem. Instead, the issue now (in contrast to the concerns of the 1970s) is that inadvertant and potentially hazardous environmental effects[4] of energy use constrain the ways in which energy *can* be used.

"Human-induced climate change can have profound consequences for the world's social, economic and natural systems."

"Although substantial scientific uncertainty remains concerning the precise time, location and nature of particular impacts, it is inevitable, under the scenario developed by Working Group I, that in the absence of major preventive and adaptive actions by humanity, significant and potentially disruptive changes in the earth's environment will occur."

Figure 1. Consensus is growing with respect to the need to avoid certain environmental futures. (Cf. Intergovernmental Panel on Climate Change, Policymakers Summary of the Potential Impacts of Climate Change; Report from Working Group II to IPCC; June 1990.)

Analyses of the functioning and vulnerability of such life-supporting mechanisms that underlie nutrient turnover, climatic patterns, and biological productivity now abound. Changing qualities of the various *subsystems* of the Earth (soil turnover of nutrients, atmospheric radiation properties, temperature sensitivity of the cryosphere, distribution of species and its sensitivity to large-scale changes in chemical and physical factors, etc.) are being reported. Scientific controversies on the salient characteristics of the overall *global ecological system* [5] may be the first signs of paradigmatic shifts towards new implications of the well-worn concept of interdisciplinarity. Such shifts, should they gain momentum, may well bear heavily on strategies for natural-resources and energy-resources management over the next few decades.

In addition to the redefinition and discovery of new *environmental constraints,* a number of technological, cultural, institutional, and other factors are now being identified as determinants of the *societal opportunities* for actually realizing some of the futures available within the (dynamic) stability domains[6] of the biosphere. However, the growing awareness of environmental constraints to human activities (essentially a biological-sciences perspective) is by no means linked with a consensus with respect to policy measures capable of promoting development towards environmentally *and* culturally benign futures. Nor is there consensus with regard to the salient qualities of transition processes (such as from nuclear-based or fossil-fuel-based energy systems to solar-based resource systems) towards such futures.[7]
This lack of consensus is due to a wide range of more or less tangible factors, some of which can be subjected to quantitative analyses (the world of data), while others, at best, can be qualitatively investigated (the world of perceptions). In particular, confrontations between the world of data and the world of perceptions take place across several interfaces: where science has to link with policy;[8] where hard science meets soft science;[9] where strongly technology-oriented futures scenarios have to come to grips with (or are, in fact, strongly prejudiced by implicit) deeply rooted attitudes towards Nature;[10] where the role of scientific intuition (fingerspitzengefühl)[11] has to be compared with the role of vested interests;[12] and so forth.

As an example, there are certain risk-philosophy approaches[13] that are less oriented towards traditional risk-tree analyses, cost-benefit analyses, and other conventional methods, and more towards risk as an existential phenomenon. This new perspective on risk suggests that environmental surprise could be generated that human societies all over the world for a variety of reasons would be unprepared, or unable, to cope with. Accepting this premise, human and societal abilities of safe-fail design[14] of energy and resource paths will have to be emphasized. A vain belief in the possibilities for designing fail-safe paths cannot match the wide variety of major, and increasing, uncertainties typical of global change issues. That is, the paradoxes of coping with surprise[15] now move (or should move) towards the top of the policy agenda, involving needs for advancing our risk perceptions *vis-à-vis* natural systems.

To sum up, there are a number of *environmental* constraints and opportunities that determine, within given time-frames, a set of global, regional and local futures available[16] to man. However, there are also *societal* contraints and opportunities for implementing paths towards realizing environmentally and culturally sustainable futures. Particular, and crucial, facets here are the linkages between energy-policy *formulations* in the global-warming era[17] and *implementation* of measures (be they based on limiting expected change, adapting to change should it occur, or on combinations of limitation and adaptation). For instance, the manner in which the Brundtland Commission[18] concept of sustainable development[19] can be defined, and which indicators of sustainable development can be chosen and applied, will strongly bear upon our abilities to reconcile the potentially conflicting goals of achieving global as well as local sustainability.

Formalizing the energy equation – a framework for discussion

Although the energy sector is the single most important contributor to the problem of global warming the importance of a variety of other sectors must not escape notice. Virtually all major societal activities are involved in creating the problem, and must become involved in solving it. Scientifically, the tasks are interdisciplinary.

To deliberately carry to its extreme the confrontation between an engineering thought process and a social-sciences approach, a formalized equation is introduced.

For interdisciplinary discussions of some of the above confrontations between the world of data and the world of perceptions a formalized frame-work may be helpful. First, rather than taking the usual perspective where contributions to the anthropogenic part of the greenhouse effect are being related[20] to radiatively active trace substances[21] (the so-called warming pie),[22] contributions to the warming pie can be reaggregated to estimate their origin in various societal sectors [23] of the global system (Fig. 2).

Unfortunately, the nation-by-nation details of the warming pie, seen from this sector-oriented perspective, are not normally available yet. For instance, whereas CO_2-data from Sweden are available, little is known about gross or net emissions of the remaining greenhouse gases. However, a rough exercise[24] for Sweden results in the estimates given in Fig. 3. Clearly, examples of even higher degree of warming-pie resolution, for instance on county-scale levels, are much needed. Such detailed analyses would help make national policies consistent with sub-

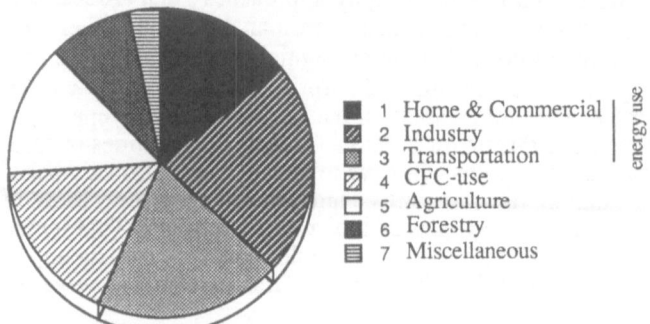

Figure 2. The global warming pie. Sources of the greenhouse gases by sector, taking all the major greenhouse gases into account. (Cf. Lashof D.A. and Tirpak D.A. (eds.) (1989) *Policy Options for Stabilizing Global Climate*, U.S. Environment Protection Agency, Washington D.C.; Trexler M.C., Mintzer I.M. and Moomaw W.R. (1990) Global warming: an assessment of its scientific basis, its likely impacts, and potential response strategies. Background Paper no. 6 to the *Workshop on the Economics of Sustainable Development*, Washington, DC, January 23-26, 1990; World Resources Institute (1990) *A Guide to the Global Environment*, 1990-91. Somewhat different data are given by IPCC Policymakers Summary of the Formulation of Response Strategies, Report prepared for IPCC by Working Group III, June 1990. Estimates are associated with large uncertainties.)

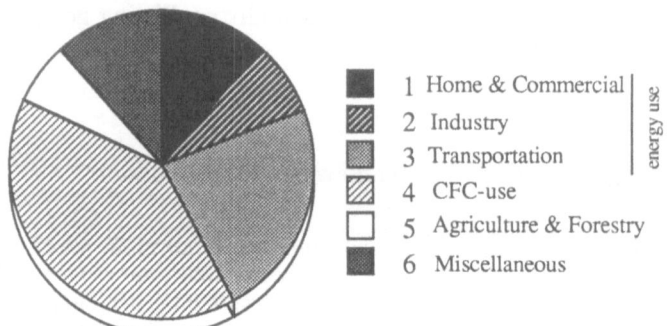

Figure 3. The Swedish warming pie. Sources of the greenhouse gases by sector in Sweden, taking major greenhouse gases into account. (Based on currently available emission data. Estimates are associated with large uncertainties. Cf. also Fig. 9 for Forestry. See Appendix for details.)

national goals through clarifying which sub-national regions that are net sources or net sinks for greenhouse gases. Such souce-sink details could indicate possibilities that are currently being overlooked for optimizing trade-offs between such regions in a way that would reduce overall national emissions.

Figs. 2 and 3 show the well-known fact that the energy sector is the single most important contributor to the problem of global warming, mainly through CO_2-emissions. However, the importance of the complexity of the warming pie must not escape notice, a feature also obvious in the estimated greenhouse-gas contributions from Swedish sectors.

Firstly, virtually all major societal activities are involved, not in the least forestry, agriculture and waste management in creating the problem.

Secondly, not only can each such sector be a part of solving the problem, but clever interplay *between* sectors could add to the range of policy options. That is, managing the problem through extending policy options, *and scrutinizing their interactions,* will be as important as refining technical responses within one sector (say, increasing efficiency within the energy-use sector,[25] or introducing environmentally less hazardous substitutes for freons and halons [26]).

With the potentials for inter-sector interactive policies in mind, now focus on energy use, and summarize the logic. Whether or not the anthropogenic signal of global warming and climate change is considered to exist, assume that risk perceptions *vis-à-vis* natural systems lead to accepting that scientific uncertainties cut both ways, and hence, that there *can* be unexpected runaway behaviour (i.e., such rates of climate change that society cannot respond adaptively to). The policy problem (P), then, inevitably is one of risk-minimizing (or at least risk reduction). The task is to envisage solutions (S) for minimizing societal and environmental risk.

The analysis can, in principle, follow either a nuclear-future perception or a solar-future perception. To simplify the following discussion I will assume that the nuclear-future option is less interesting on the global scale and that the solar-future option merits critical analysis.[27]

Major solar-future options identified hitherto for approaching a solution to sustainable futures are based on increasing energy-efficiency (efficient use of energy, EUE) in combination with phase-in of renewable energy sources (RES) and phase-out of sources identified as either being too far from technological break-through,[28] or being environmentally unsound, or societally unmanageable or unacceptable.[29] That is, to deliberately carry to its extreme the confrontation between an engineering thought process and a social-sciences approach, a formalized "equation can" be set up:

$$EUE+RES \rightarrow Solution?$$

Those energy sources and options that underlie the existence of the problem (P) do so through not complying with a set of boundary conditions. Alternative sources and options also face, of course, a set of boundary conditions. In order to imply any reduction of risks these conditions must be complied with. Some of the conditions are (several aspects of) cultural acceptability; environmentally benign qualities; security and control constraints; capability for easing (geo)political tensions; properties appropriate for managing debt crisis; and qualities conducive to technology exchange, transfer, leapfrogging, and integration into existing infrastructures.

That is, for solutions S to be viable they have to be carefully screened for inconsistencies with respect to boundary conditions. These will, in fact, be partly normative (prescriptive), partly geophysical (for instance, based on some acceptable degree of scientific consensus [30]), and partly technological (for instance, based on some acceptable degree of uncertainty with respect to the commercializing of technologies only demonstrated under laboratory conditions). But solutions also have to be investigated from aspects of interactions between the energy sector and other sectors; that is, there is *also* a set of what could be called "opportunities for relaxing boundary conditions".

Problem P
risk minimizing
Option
efficient use of energy
Option
renewable energy sources
Task
approach a solution within time t<ts
(ts is year when solution has to be fully implemented to be meaningful)
Equation
$$EUE(t) + RES(t) -> S(t), t=time$$
Boundary conditions
data [A] and *perceptions* [B]
Opportunities for relaxing boundary conditions
data [C] and *perceptions* [D]

Figure 4. Formalizing the "energy equation" for further discussion. ([A]E.g., geophysical, such as satellite-based tropospheric temperature changes. [B]E.g., acceptability of large-scale ecological engineering of carbon-sinks and/or energy plantations. [C]E.g., for reducing methane emissions from the waste-management sector and at the same time, through channelling methane through the energy system, generate a spin-off in the energy or transportation sector. [D]E.g., through promoting unconventional measures for reducing obstacles in the process of linking science with policy, or in the process of harnessing green market forces, or with respect to promoting creative thinking about alternatives (or complements) to GNP-measures, tracking, for instance, national energy and resource efficiency.)

Finally, the energy equation is dynamic in several respects. For instance, the credibility of some renewable resources will depend on the rate at which energy-efficient technology permeates the energy system. If suitable niches for (the conceivably medium-utility or mini-utility scale) renewable sources open up, this, in turn, would tend to stimulate the development and implementation of energy-efficient tapping of the renewable sources.[31]

Also, the perceived time-scale of the problem will bear upon (to some degree even drive) the time-profile for the emerging solution. The energy equation is formalized into a flow-scheme in Fig. 4.

A brief discussion now follows of the various elements of the equation, i.e. risk perception, efficient use of energy, renewable energy sources, boundary conditions, opportunities for relaxing boundary conditions. The intention is to try to address potential conflicts that may reside within the solar-future option, and to attempt to outline ways to resolve such conflicts. Each element of the formalized equation can then be commented on from the two alternative perspectives introduced at the outset of this paper: perceptions versus data. Note that major portions of the comments on data draw on the Appendix.

In commenting, my perspective is that the existence of technologies for approaching solutions to the equation, and the cultural and ecological analysis, must come prior to discussions on economic potentials and constraints. Therefore, conventional economic aspects are beyond the scope of this paper.

Risk minimizing – facing Nature Non-Linear

Natural systems are diverse organizations, with many components that are strongly connected. That is, natural systems belong to the class of complex systems. What does this complexity mean – a safeguard or a trap-door?

Perceptions

The possibly on-going shift as to the perceptions of Nature's *modus operandi* is currently being discussed at some length from the humanist's perspective[32] and from the systems theory perspective.[33] Similar discussions are presented in cultural anthropology.[34] From an ecological theory viewpoint a background to this discussion can be found in a paper by the renowned ecologist Eugene P. Odum in the *Japanese Journal of Ecology* in 1962,[35] laying some of the essential foundations to scientific conceptions of the functioning of ecological systems. This paper by E.P. Odum, and many other papers by well-known ecologists, seem to have (quite unintentionally) surfaced on the environmental policy arena as a policy paradigm of Nature Benign, or Nature Linear (cf. Fig. 5).[36]

These paradigms seem to contribute much to the business-as-usual (or wait-and-see) attitudes that exist in the policy debate on the societal response to scientific information on global warming.

The new paradigm, potentially evolving, stresses a quite different perspective: Nature Non-Linear (or Nature Complex). Whereas the Nature Benign view emphasizes the existence of stabilizing (homeostatic, geophysiological, gaian [37]) properties of natural systems, the Nature Non-Linear school of thought highlights the potential that systems complexity implies destabilizing properties, rather than stabilizing ones. Such properties can be generated by non-linear dependencies between system components, conducive to flip-flop, threshold, or run away tendencies[38]. That is, to simplify for brevity, the complexity of natural systems (in particular the global ecological system) is seen either as a safe-guard against anthropogenic stress or as a trap-door that can change, surprisingly, the *modus operandi* of natural systems. These extremes lead to very different environmental policies (cf. Figure 5).

The risk concepts that are evolving differ from classical risk-tree analysis, and related approaches. The question involved essentially relates to the existence and extent of a dynamic biospheric stability domain, for instance with respect to the rate of average global warming.

Clearly, a Nature Non-Linear approach to risk minimizing would lead not to wait-and-see strategies, but to a strategy aiming at reducing the rate of change, as well as the absolute change, imposed by humankind on the biosphere.

Data

There are data to support the idea of Nature Non-Linear; for instance, relating to the delayed discovery of the unprecedented spring-time depletion of the Antarctic ozone column, and also to the occurrence of "Waldsterben" in central Europe.[39] It will be mandatory, in any risk assessment, to thoroughly consider the interplay

Paradigm:
NATURE BENIGN
complexity a safeguard
against environmental response to stress
Nature infinitely forgiving

Policy:
BUSINESS AS USUAL

Paradigm:
NATURE LINEAR
complexity a safeguard
against irreversible response to stress
Nature signalling response

Policy:
WAIT AND SEE

Paradigm:
NATURE NON-LINEAR
complexity absorbs stress
hides response
then generates surprise
Nature giving early warning?

↓

Policy:
INSURANCE BUYING

Figure 5. Some perspectives ("half-truths", "myths") underlying environmental and natural-resources policies. *Nature Benign* can be considered out of date; *Nature Linear* seems reasonable when studying stress-effect relationships within the stability domain. Outside of this domain, *Nature Non-Linear* (or *Nature Complex*) constitutes a world of perceptions which now has to be developed into operational tools.(From Wiman B.L.B., Implications of environmental complexity for science and policy, *Global Environmental Change,* 1991, accepted)

between such traditional and already experienced environmental pertubations and global warming.

For instance, increasing temperatures may lead to increasing mercury emissions from soils and increasing stomatal uptake of mercury in forests. This would result in a kind of inadvertant redesign of the biospheric chemistry of mercury (shifting from one pool – soils – to another – biomass).[40]

Of particular interest for the relationships between energy and the (global) environment would be a reliable quantification of the temperature sensitivity of the biospheric stability domain (assuming now that risk minimizing is to comply with the Nature Non-Linear school of thought). The often quoted value of a maximum allowable rate of average temperature change of about 0.1 °C per decade[41] (and an absolute increase of 1 or perhaps 2 °C over the next hundred years) can be seen as a first approximation of the problem of defining the stability domain. For several reasons one may recommend a much closer look at this problem.

One reason is that little is known about the distribution over time of such a value: for instance, would 0.4 °C be allowable over one decade if compensated for by four consecutive decades of only 0.025 °C per decade?

Another reason is that such values so far seem to be based on paleoclimatological data; that is, data pertaining to a quite different biosphere than today's. For today's biosphere there is the essential need to enable a better quantitative and qualitative understanding of the implications of the about 0.5 °C average global temperature change over the last century, i.e., in the industrializing biosphere era.

As is well known, however, the existence of a *suggestion* for a rate-of-change criterion has triggered a set of policy formulations to which I will return later.

To sum up, the complexity of natural systems can *either* imply a safety net with respect to anthropogenic impacts on the climate (the Nature Benign – or complexity as a safe-guard – policy paradigm) *or* that complexity paves the way for unmanageable surprise (Nature Non-Linear – or complexity as a trap door – policy paradigm). As there is yet no clear understanding of which paradigm that is valid, there is an obvious need for decelerating the rate of global warming, regardless of whether or not there is data enough to support the standpoint that climate change has already been triggered by man.

The costs of adapting to a non-linear, unpredictably galloping climate change cannot be calculated, and could prove to be insurmountable, whereas the costs of limiting change do seem to be possible to at least estimate.

The risk-philosophy situation does have some traits in common with other societal and institutional insurance-buying aiming at reducing risk.[42]

I now turn to brief comments on energy efficiency and renewables as options towards complying with such a need for decelerating the rate of change.

Option: the efficient use of energy

Indisputably, perceptions and data are changing with respect to the functioning and tasks of the societal energy systems. Top-down approaches are being complemented with (or even abandoned in favour of) bottom-up analyses.[43] The role of information technology (minor amounts of energy to channel major flows effi-

ciently) may offer important opportunities for achieving high levels of energy-supply and end-use efficiency.

Perceptions

Providing, marketing and investing in energy-efficient end use is becoming more interesting than investing in new generating capacity.[44] Concepts like "basic needs, and much more, with one kW per capita",[45] almost unthinkable 10-15 years ago, have provoked discussions on how the needs for services and goods of a society could be met through other means (such as solar-powered vehicles) than the conventional ones (such as gasoline-powered cars). In particular, interest during the 1970s (when the basic rationale was the oil crises) in the potentials for increasing the efficiency of anthropogenic energy systems has now been revitalized. The current driving force behind this renewed interest is mainly the concern about the environment and global security.

However, the question is often raised: why – as efficiency measures could even be economically superior for energy utilities as well as for the private end-user – does increasing energy efficiency permeate society so slowly?[46] Several barriers have been identified, as have several means to overcome these barriers. It is, however, beyond the scope of this paper to discuss these topics. It suffices to say that measures suggested include cost-neutral incentives for alternative energy strategies, energy-efficiency labels on appliances, and improved public information.

Data

Considerable debate is taking place with respect to what reductions in specific energy end-use (energy use for a given service or goods) could be achieved over, say, the next twenty to thirty years.[47] The data collection in Fig. 6 indicates some examples of current estimates. For instance, some processes in industry (such as grinding) are not expected to reduce their use of electricity by more than 15% (relative to current use) over the next twenty years or so, whereas others (such as more efficient lighting technology) could achieve 55% reductions, or more. Improved fuel economy of cars, and use of non-conventional fuels, could result in specific energy-use reductions of around 75% in the transportation sector.

An overall efficiency improvement for energy systems of around 30% to 50%,[48] and even of 50% to 75%[49] over the next twenty to thirty years has been suggested. Broadly speaking, the 30% improvement level (corresponding to an average improvement of efficiency of around 1% per year) seems to be consistent with the rate of improvement experienced over the last few decades and might therefore be considered a value pertinent to what could be called normal turnover-rates of technology. The 50%-improvement level (around 1.7% per year) is considered by Bodlund et al. (1989)[50] as attainable for Sweden, given policies are shaped to promote individual actors' interest in benefitting from the increased rate in efficiency improvements.

A 75%-improvement scenario (around 2.5% per year), while within reach, might have to be linked with major shifts in attitudes and policies. Although there do not yet seem to be comprehensive data available, the role of information technology (minor amounts of energy to channel major flows efficiently) is likely to become an integrated part of paths towards high levels of efficiency improvements.

It may be noted that although scenarios building on increased energy-efficiency for Sweden indicate essential keys to reducing CO2-emissions, the supply and end-use

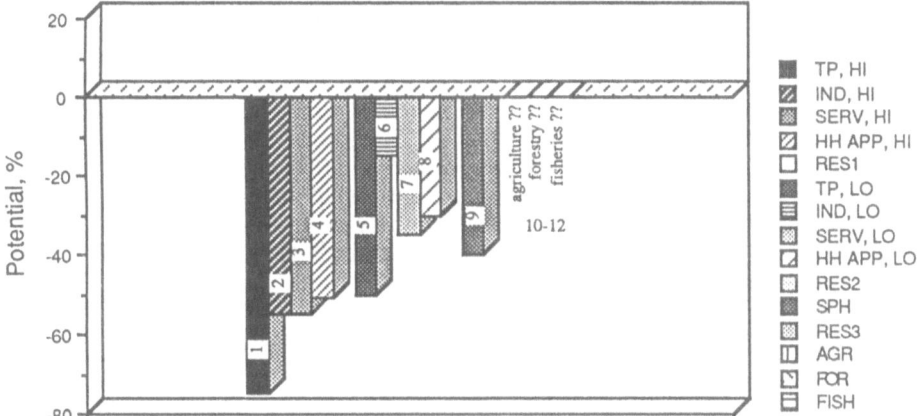

Figure 6. Potential for specific energy end-use in around 2020 relative to 1990; some examples. Resulting overall scenarios for nation-specific improvements depend on several factors, including the assumed end-use mix. (Based on Bodlund B., Mills E., Karlsson T. and Johansson T.B. (1989) The challenge of choices: technology options for the Swedish electricity sector. In T.B. Johansson, B. Bodlund and R.H. Williams (eds.) (1989) *Electricity. Efficient End-Use and New Generation Technologies, and Their Planning Implications.* Lund University Press, Lund, pp. 883-947. See Appendix for details.)

Legend: 1 – transportation, high estimate; 2 – industry, high estimate (e.g., for improved lighting efficiency); 3 – service sector, high estimate (e.g., for lighting technology); 4 – house-hold appliances, high estimate (e.g., more efficient refrigerators); 5 – transportation, low estimate; 6 – industry, low estimate (e.g., energy for grinding processes); 7 – service sector, low estimate; 8 – house-hold appliances, low estimate (e.g., TV equipment etc.); 9 – space-heating (e.g., through exploiting co-generation capacity, district-heating, etc.); 10-12 – agriculture, forestry, fisheries - no data.

mix in the energy system, and uncertain factors such as attitudes and expectations, are also major determinants in shaping greenhouse-gas policies.

Option: renewable sources for energy – and for carbon sequestering?

Increasing efficiency is an essential factor in managing the energy equation. But current, or anticipated, technology for efficient end-use cannot alone provide solutions. Low-carbon supply strategies are necessary as well as capabilities for withdrawing excess carbon already residing in the atmosphere.[51] What can solar technology and techno-ecological engineering contribute?

Perceptions

Similar to the interest of the 1970s in energy efficiency (and for quite the same reasons), and its revitalization over the last years, renewable energies are now clearly on the environmental agenda again. Several observers, not in the least in Sweden, have noticed the slow progress, however, in implementing (at least on the large scale) energy systems based on renewable sources.

It would take this paper much too far to go into details with respect to the Swedish debate on these issues. However, parts of the debate have centered on the case for artificially low pricing of fossil-based and, in particular, nuclear-based energy supply. The pricing strategy, linked to a surplus of electricity, may therefore have put renewables into an unnecessarily disadvantageous position. At present, Sweden faces three important challenges for the future of the energy systems: a nuclear phase-out, a recommended freezing of carbon-dioxide emissions at about 1988 levels, and no tapping of the four remaining unexploited large Swedish rivers. There are signs that the political process towards consensus on the time-profile for the nuclear phase-out will be embedded in agreeing upon a time-profile for the phase-in of renewable sources.[52]

However, widening the perspective from Swedish specifics to more general principles, another set of concerns could also underlie – or could come to underlie – the shift towards extensive use of renewable energy sources. Some of these concerns were expressed already during the renewables debate in the 1970s, although by no means as eloquently as now seems to be the case. The point is that renewable energy sources (rightfully or not) are associated with the perception that very large areas must be transformed into energy sources. This perception invokes a wide range of more or less intangible questions. Some are obvious, pertaining to conflicting needs regarding large (more or less) fertile areas (food versus fuel, recreation versus unpleasant energy plantations, etc.). Others belong to a class that could be called "deeply rooted attitudes towards Nature".
To return for a moment to Sweden, the concept of culture, agriculture and silviculture being strongly intertwined has profound roots. This situation certainly has analogies in other countries. These cultural facets are now being confronted (cf. Fig. 7) by large-scale, even global, environmental challenges not at all easy to tie together into a comprehensive view on the role of renewable energy sources. The debate on biotechnology (sometimes assumed to have a leading role in implementing biologically engineered biomass-systems) adds to (or will probably add to) the issue of attitudes towards "engineering Nature". Further, as exemplifed by Wright (1990)[53], increasing manmade diversions of energy flows from natural systems to human systems could result in massive species extinction. If so, major potentials for discovering, and using, species fundamental to living and future human generations would be undermined.[54]

Finally, at least as far as biomass-energy is concerned, not only are there seemingly conflicting goals between food production and energy supply, but also a growing debate on the use of large-scale ecological engineering of large land areas turning them into tactical or strategical carbon sinks.[55] From a natural-sciences perspective, several important aspects need to be clarified with respect to the biogeophysical and biogeochemical implications of such large-scale sinks (for instance, their release of terpenes and isoprenes, and thus their effect upon regional atmospheric aerosols).[56]

In conclusion: adding to traditional obstacles to the introduction of (efficiently used) solar-energy technology (such as establishing techno-economic evolutionary paths and niches [57]), solar-engineering and carbon-sequestering technology (through land-area demands) links in particular ways to attitudes towards Nature. This introduces particular needs to engage scholars from the humanities in solar-technology R,D&D. This is because, as will be commented on in the next section, the high technological potential and non-polluting qualities of solar energy will not be sufficient prerequisites for a reliable solar option for the future. To be regarded as a fully viable option also from cultural aspects, solar engineering must be

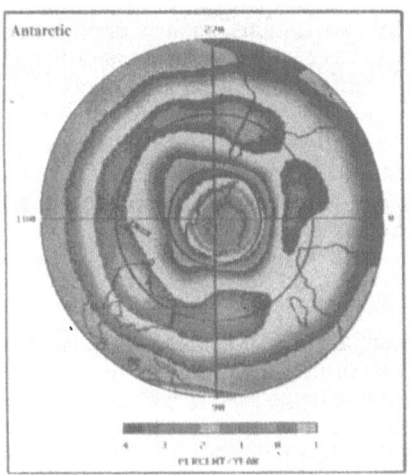
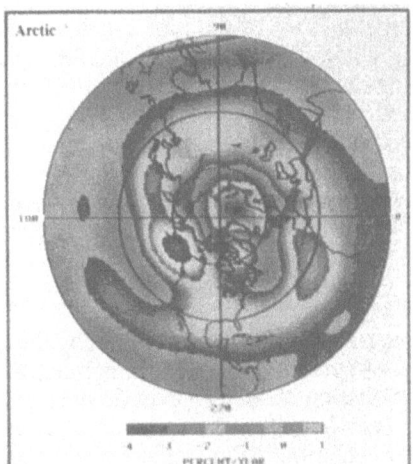

Figure 7. Clashing perspectives? Below: a deeply rooted Swedish (folk-lore) perception of Nature as benign and fully sustainable. Above: the ozone-depletion effect illustrated for the Antarctic (left) and the Arctic (right).(Montage from Wiman I.M.B. (1988) Natursyn i förändring [Changing perceptions of Nature]. *Bygd och Natur, Årsbok 1988*, 15-30 (picture given in colour in that reference; ozone-depletion as illustrated in *New Scientist* 6 November 1986, p. 37).

designed to fit into region-specific traditions *vis-à-vis* more or less intangible qualities of Nature. In fact, a long-lasting dialogue in ideo-historical research on the idea of "the Earth as a garden" will inevitably have to become involved in large-scale solar engineering – a seemingly bizarre but nevertheless essential rendezvous between technology, natural sciences, and the humanities.[58] As one example, carefully designed agroforestry systems [59] – in particular such applications that build on the vast possibilities of using non-conventional species[60] – could contribute to solving dilemmas and easing tensions involved.[61]

One may say that a "renewables romanticism era" of the 1970s is now turning into a "renewables credibility era". However, large-scale landscape redesign should not forget lessons of the past. Ancient western civilizations, such as the Romans, fearing wilderness and unpredictability, seem to have been able to perceive Nature as being in proper shape only if human control and manipulation were everywhere present.[62]

In ideo-historical research this attitude has been commented on by several scholars, for instance: "The Romans' failure to adapt their society and economy to the natural environment in harmonious ways is one of the causes of the decline and fall of the Roman Empire, if not in fact the basic underlying one."[63]

Data

There is no indisputable way whereby the potentials of various renewable sources can be compared, or ranked, in compatible measures. However, one feature that the exploitation of such sources do have in common is the need to collect the rather thinly distributed – albeit essentially abundant, cost-free, and inexhaustible – supply of solar radiation incident on the surface of the Earth. The Earth-atmosphere system receives energy from the sun at the prodigious – but only deceptively large – rate of close to 172 000 TWyear per year, whereas the total primary energy use in all man-made energy-generating systems in the world in 1970 was less than 0.004% of this (about 6.9 TW), and in 1990 is about 0.007% (about 12 TW).

A common denominator for all solar-based energy systems is thus a need to match the thin distribution of solar energy with surface areas large enough to collect significant amounts of radiation. This leads to a kW per unit area measure (i.e., kW years per year and hectare) of the potential for various solar technologies. The area of one hectare will be used in the following to indicate a level-of-scale pertinent especially to biological (or ecological) solar engineering.

Figure 8 (based on more detailed considerations in the Appendix) exemplifies such measures, and also indicates some of the deceptive character of the vast quantities of solar energy that seem available for harnessing.

For instance, a solar-cell panel in a suitable orbit in space would be capable of an electricity output of around 2000 kW per hectare, but a realistic value for a similar system on the Earth's surface would only be around 200 kW per hectare (of plant site).

Wind- and wave-energy systems fall in the class around 30 kW per hectare (although part of the hectare could be used for other activities, such as cropping, or aquaculture); whereas energy-crop plantations could provide energy corresponding to about 5 kW of electricy per hectare (again, multiple-use design measures could add to the quality of this hectare).[64]

With these constraints and potentials in mind, it is interesting to note that the total primary-energy use in Sweden's (anthropogenic) energy system is about 1.3 kW per hectare, about the same as used in Africa (averaged over the total area of the continent). One major difference between Sweden (with about 0.15% of the world population) and Africa (with about 12.2% of the world population) is that the use of domestic energy sources in the energy system is around 14 % in Sweden, but around 75% in Africa.[65] At the same time, the forests of Sweden (currently in a state of net growth) add a carbon-sink capacity to the Swedish territory (cf. Fig. 9), whereas the forests of Africa are diminishing (with environmentally unfortunate effects such as unclear contributions of carbon to the atmosphere). On the other hand, the yearly output of fossil-fuel carbon dioxide (and other greenhouse gases) per capita from Sweden is many times (one order of magnitude greater than) the

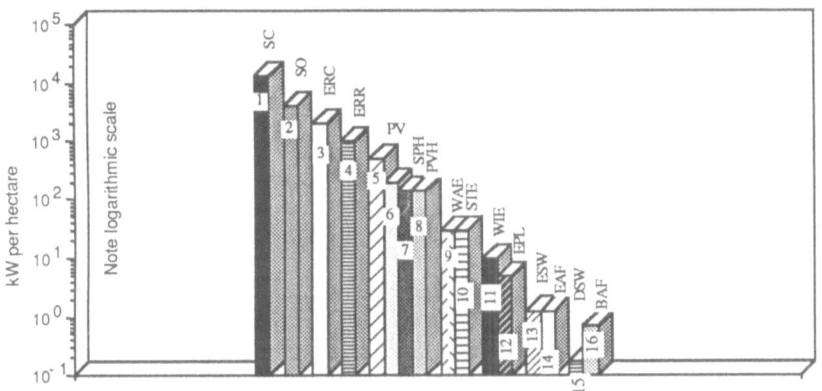

Figure 8. Solar energy data (kWyear per year and hectare) for various systems. (See Appendix for details.) Note the seemingly enormous potential available if current technologies and uses are compared with the solar constant. A wide variety of constraints (physical, such as annually received energy at a given location being determined by cloudiness and other optical depth parameters; technological, such as solar-cell efficiency or energy inputs for harvesting and transportation of energy forests) determine the actual amounts available. Also note (see text for further details) that if a range of non-technical issues were to be considered, the ranking may well look very different (for instance, wave-energy systems could accommodate aquaculture, adding important aspects of multiple-use to the ranking).

Legend: 1 – solar constant; 2 – electricity output from solar-cell panel in suitable orbit around the Earth; 3 – energy received at the Earth's surface, clear conditions; 4 – energy received at the Earth's surface, average conditions (clouds, etc.); 5 – electricity output from PV-panel, 28% efficiency, located in the tropics, clear conditions; 6 – electricity output from PV-power site (kW per hectare of site area, 10% cell efficiency); 7 – solar space-heating system (equivalent electricity output); 8 – useful energy output from PV-hydrogen system; 9 – electricity output from wave-energy system; 10 – output from solar-thermal electric system; 11 – electricity output from wind-energy system; 12 – equivalent electricity output from energy plantations; 13 – total primary-energy use per unit area in Sweden; 14 – total primary energy-use in Africa per unit of total area of the continent; 15 – primary energy-use per unit area in Sweden emanating from domestic fuels; 16 – primary energy-use per unit area in Africa emanating from biomass.

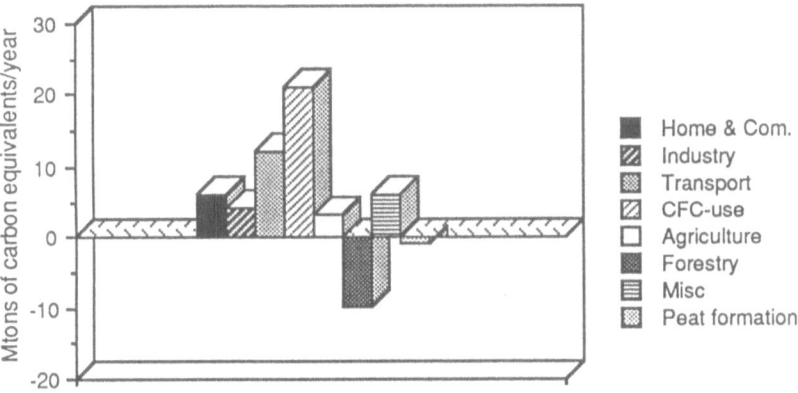

Figure 9. Equivalent-carbon source/sink strengths in Sweden. (Estimated from currently available data; see Appendix for details.)

corresponding value for Africa. In turn, Sweden releases less than half as much carbon dioxide per capita and year than do the United States.

The obviously and severely disproportionate patterns exemplified above call attention to a wide range of national and international issues. For brevity, the questions are captured by the phrases "think globally – act locally" and "think globally – act globally". For instance, should Sweden not only reduce national carbon dioxide emissions but also engage in tactical carbon-sink plantations abroad? Solar cells in Sweden, and efficient use of energy, could imply that certain carbon emissions could be avoided.

But neither solar cells nor energy-efficiency can withdraw carbon dioxide already in the atmosphere.

In Fig. 10, the potential carbon-sink (or carbon-release) capacity of various species and systems is exemplified. As has been noted by Houghton (1990)[66] it could be possible, from a pure sink-source point of view, to use vegetation (through biomass-energy systems in combination with carbon sequestering) to eliminate the total net release of carbon to the atmosphere from fossil-fuel combustion and from deforestation. Other approaches, such as that proposed by Firor (1988)[67] lead to similar conclusions.

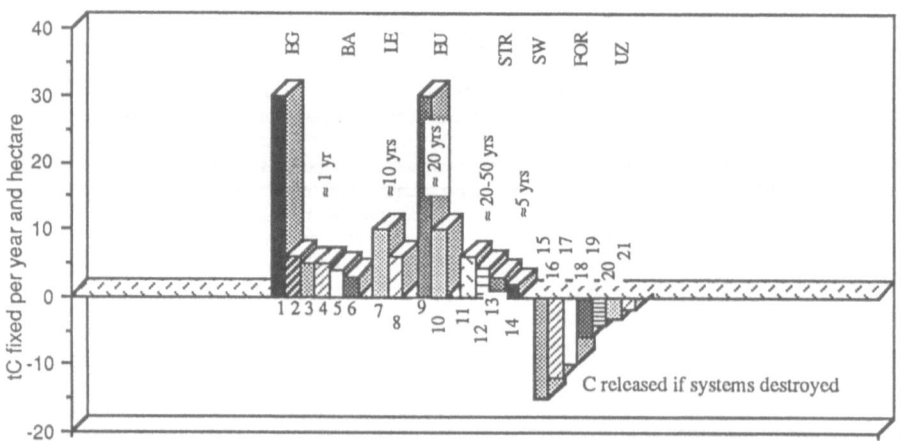

Figure 10. Potential carbon sink (or release) capacity of some species and systems, estimated representative values. (See Appendix for details.) The indicated time-horizons pertain to the thought experiment that the species (or systems) were used to sequester carbon. That is, to use potatoes, for instance, not for food but for sinking atmospheric carbon, every year's harvest would have to be withdrawn sufficiently long from the carbon cycle, whereas a strategical carbon-sink plantation in a state of net growth would contribute to sequestering carbon over about 20 to 50 years. This thought experiment serves to highlight conflicts with respect to land-use for food, fibre, energy plantations, and carbon sequestering. See main text for further discussion on means to resolve conflicts.

Legend: 1 – elephant grass; 2 – potato; 3 – corn; 4 – reeds; 5 – lucerne (alfalfa); 6 – barley; 7 – Leucaena; 8 – willow, poplar; 9 – Eucalyptus; 10 – Eucalyptus, low estimate; 11 – strategical (temperate or tropical) carbon-sink plantation; 12 – tropical plantation, medium age; 13 – tropical plantation, mature; 14 – tropical plantation, early stage; 15 – Swamps, marshes; 16 – algal reefs; 17 – tropical rainforest, natural; 18 – temperate forest; 19 – savannah; 20 – temperate grasslands; 21 – upwelling zones.

Strategies based only on ecological engineering will require very large areas, however, aggravating conflicts between various users. Including reforestation of tropical ecosystems, around 500 millions of hectares are needed, corresponding to around 30% of the world area of croplands, or to around 25% of the area of tropical forests. As noted by Houghton: "The prospect of managing such areas of forests sustainably is sobering." 500 millions of hectare equate about 13 times the land area of Sweden or Japan.

Combining the above potentials for solar-energy technology and carbon-sequestering with the potential for energy-efficient measures (many of which are commercially existing) would then provide a "technical existence" proof for energy futures with little, or even fully eliminated, impacts on the carbon dioxide concentrations of the atmosphere.

Again, one may emphasize that careful scrutinizing of cultural as well as biogeo-physical and biogeochemical aspects of such very large-scale engineering redesign of the Earth's surface will be mandatory. But if carefully implemented, the principle above does suggest that a path towards reduced-risk futures could be phased in as the world's (in particular, the industrialized world's) heavy reliance on high-risk fossil and nuclear systems is phased out. As indicated in Figure 8, solar technology based on PV-cells (possibly augmented by PV-hydrogen systems) and solar-thermal systems can provide additional and very important solutions to limiting the conflicts between areal management and uses. This would be particularly true if PV-systems, for instance, could exploit the opportunities for integrated design such as in building-architecture (roofs and walls).[68]

Even better would be to exploit a variety of opportunities for integrated vegetation-buildings-PV designs. Although urban summer heat islands to some extent could be mitigated by simple measures such as high-albedo roofs and walls, more could be achieved through using a combination of PV-roofs, lighter-coloured city surfaces, and urban tree plantations.[69]

Boundary conditions – and ways to relax them: perceptions and data

The theme of energy and development looks very different from a global perspective as compared to a local perspective. As a result, the task of finding global-level indicators of sustainable development differs completely from the task of identifying local-level indicators. Whereas global-scale indicators are likely to take the form of metric measures local-scale indicators will inevitably be much more complex.

Safe-failing the energy equation is equivalent to reconciling local and global design of resource systems so as to make unpleasant surprises – when they inevitably occur – of a manageable scale.

Some of the existing constraints and possibilities for managing the energy equation through combinations of efficient use of energy and renewable sources have already – more or less explicitly – been commented on above.

It could be instructive to use a somewhat different perspective to sum up some of these constraints and possibilities. In Fig. 11 a "top-down" perspective on natural-resources (and thus also energy-resources) management is illustrated, based on the concepts advanced by Krause *et al.* (1990).[70] That is, assume it were possible to assign a critical value (with some analogies to the concept of "critical load", common in the acidification-abatement policy arena [71]) for the (dynamic) biospheric stability domain. Scientific consensus with respect to such a critical value would then be conducive to a process where allowable temperature-changes per unit time (and allowable maximum temperature-change over a wide time span) would define policy goals.

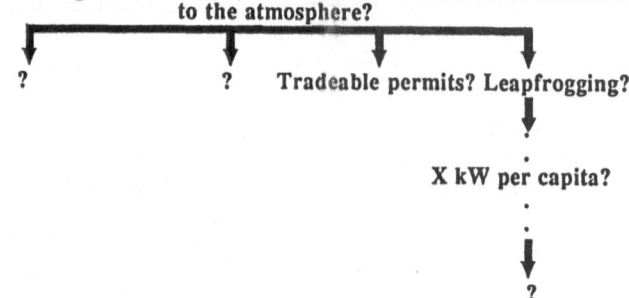

Biospheric stability domain <0.x °C per decade?

Base energy strategies and policies on a budget of 300 Gt for cumulative fossil carbon releases between now and the year 2100?

Return carbon storage in land biota, and eliminate all net biotic carbon releases to the atmosphere?

?   ?   Tradeable permits? Leapfrogging?

X kW per capita?

?

Figure 11. "Top down" in natural-resources management? Assuming a biospheric stability domain can be identified this would be conducive to a process where allowable temperature changes per unit time would define policy goals. Allowable temperature-change per unit time may well be a function of time itself. The policy process could be expected to include various forms of trading (or swapping) with respect to emission permits. Indicators of progress on the global scale would take the form of experienced temperature changes; on the regional-local scale efficiency in energy-use and resource-use could indicate aspects of sustainability. An important question is whether this (somewhat technocratic) perspctive can meet a bottom-up perspctive, cf. Fig. 12. See main text for further discussion.

These goals, in turn, would take the shape of a time-profile for anthropogenic greenhouse-gas emissions, which, in turn, would lead to possibilities for managing fundamental problems of IC (industrialized countries) versus DC (developing countries) responsibilities in responding to risks involved in global warming. Whether or not tradeable permits, potentially geared to a system for swapping advanced and environmentally benign technologies (or ecotechnologies), could offer opportunities for an IC-DC dialogue is an open question. Further, whether or not some measure like "x kW per capita" would prove useful in discussing a reasonable level of energy-intensity in societal systems is also open to debate.

That is, the essential perceptions embedded in Fig. 11 are that (i) there are severe constraints (a maximum allowable rate of change) but also (ii) a number of possibilities (such as clever interplay between sectors to achieve the goals of greenhouse-gas emissions reductions). It is not possible to accomodate within this paper any detailed discussion of such potentials for interplay; however, a few examples can illustrate the idea.

Taking Sweden as a starting point, it seems clear that any concept of freezing carbon emissions from anthropogenic sources would be deceptive if, for instance, the rate of forest-use for pulp and paper would begin to exceed the net growth rate of the forests (cf. Fig. 9). An increased *net* emission of carbon from Sweden would result, despite a freeze on anthropogenic emission sources. Instead, increasing the area used for forestry, in combination with more efficient use of wood products and reduction of fossil-fuel based transportation intensity could provide a freezing of net carbon emissions. (For instance, increased recycling of paper products would not only lower the energy use for paper production, but would also reduce the number of trees actually used to provide specific services and goods, such as books).

Further, a much more efficient management of waste deposits could supply useful gas, and would at least lower the amounts of methane emitted to the atmosphere without the methane uselessly bypassing the energy or resource system. Channelling methane through the system could contribute to providing the same services at a lower amount of emitted carbon equivalents per service. Moreover, much could probably be done to lower the methane emissions from the Swedish agricultural sector. Contrary to the major problem with methane emissions in DCs (due to wet cropping of rice) the Swedish emissions are strongly linked to more easily quantified sources (livestock). Measures for reducing agricultural methane from Sweden would, in obvious ways, relate directly to the food-basket of Swedish families. Therefore, this example highlights such direct linkages between family life-styles and global warming that were completely undiscovered in greenhouse-policy debate of the 1970s.

Nitrous oxide emissions from agriculture (and forestry), although not well understood, may seek a solution in more carefully designed fertilization schemes. Such schemes could involve restoration of wetland capacities for dealing with nitrogen. In principle, such measures would comply with strategies aiming at reducing the rate of change with respect to traditional environmental problems (such as acidification). For parts of Europe (and other regions of the world) where the carbon sink capacity is being endangered by forest damage, such strategies for combatting acidification and conventional air pollution (oxidants, etc.) would link particularly well to steps towards decelerating the rate of global warming.

It should be noted, that *even in the absence of climate change,* refined or new strategies and technologies for safe-guarding the environment can prove economically advantageous. In particular, as exemplified above by the capability of wetlands to trap nitrogen and reduce run-off to coastal areas, services (not only products) from intact ecological systems should be considered in addressing the questions of formulating so-called green GNPs.

Reasoning along the lines suggested above – that is, exploiting the potentials inherent in an integrated natural-resources management strategy – would take different shapes in other nation-specific contexts. The point is the interplay between perceptions of risks and possibilities on the one hand, and detailed data for the contributions of various sectors to the problem, on the other.

As a contrasting perspective to the one given in Fig. 11 – clearly quite technocratically oriented in that it starts with a top-down "managing-planet-Earth" approach – consider Fig. 12 that attempts to illustrate a bottom-up aspect. In a sense, such an approach may turn out to be much more complex, partly due to its being much more down-to-earth. It may serve to illustrate the fact that the specifics of the setting (cultural attitudes, plant communities suitable for a sustained production of a certain service or goods, such as indicated by the symbol "honey-production") are fundamental determinants of the opportunities for implementing concepts such as sustainable development. As another example of the imprtance of not letting details out of sight, even if problems are large-scale, consider the use of *Eucalyptus* for

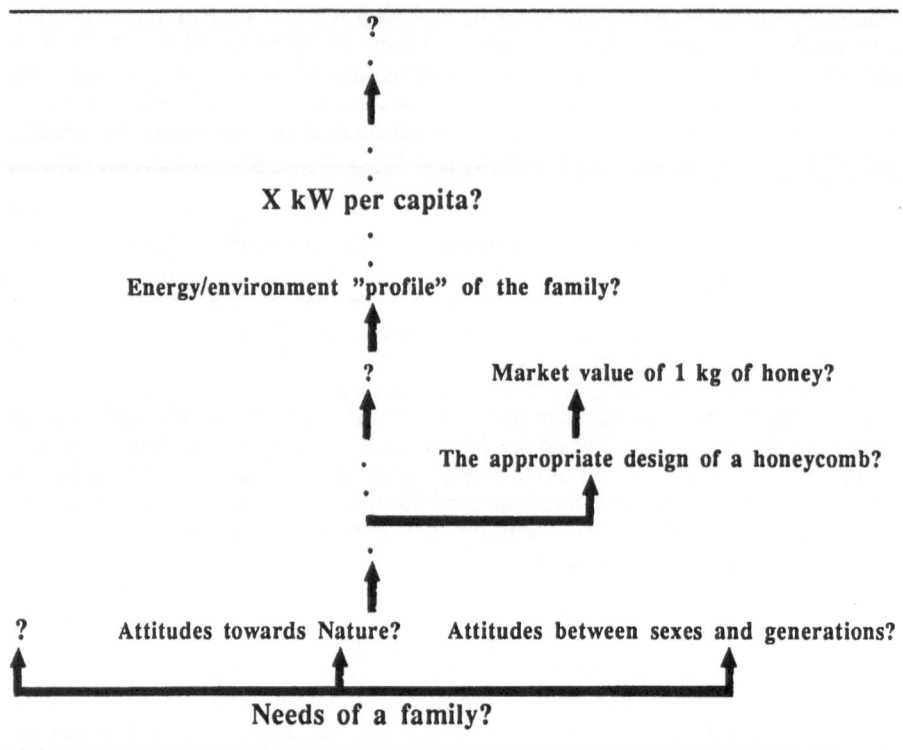

Figure 12. "Bottom-up" in natural-resources management? Starting with the needs of a family a complex web grows with respect to the meaning of sustainable development. It is not at all obvious that this perspective leads to management strategies for energy and natural resources consistent with the "top-down" approach. The "honey-comb" symbolism intends to illustrate the fact that cultural attitudes, plant communities suitable for sustained production of honey, and so forth, determine opportunities to match large-scale management (in terms of strategical carbon sinks, etc.). See main text for further discussion.

multiple-use purposes. Many *Eucalyptus* species form shallow roots competing with adjacent food species. In contrast, some species of *Leucaena* not only grow deeply rooted but are also able of providing a range of services, including fixing atmospheric nitrogen and producing a variety of interesting substances.[72]

Looking simultaneously at Fig. 11 and 12 one finds, among other things, that whereas global-scale indicators of sustainable development may well take the form of metric measures (temperature change per decade, tons of carbon emitted per capita, etc.) local-scale measures are of fundamentally different types. Clearly, there is ample reason to further analyse the implications of metaphors such as "local change – global surprise" versus "global change – local surprise".[73] By definition, our environmental futures contain surprises. Safe-failing the energy equation is equivalent to making a consistent rendezvous between local and global design of resource systems so as to make unpleasant surprises – when they inevitably occur – of a manageable scale.

Policy exercises[74] and the establishing of measures for behavioural insurance will have a role to play in that respect, as the emerging solution has to be checked and cross-checked in several ways. Not in the least, the barriers between science and policy,[75] and between the world of data and the world of perceptions, need to be exposed and wherever possible, eliminated.

One such barrier of interest, for instance to international preparations for the Brazil 1992 conference on environment and development (the follower-up to the 1972 Stockholm Conference) resides in scientific information being basically probabilistic, whereas policy decisions ordinarily require singular, discrete choices among fixed and – preferrably – mutually exclusive alternates. Another obstacle in linking science with policy is that public policy makers use scientific information within a political context.[76] One important step towards bridging the science-policy communication gap would be to promote openess and transparency with respect to underlying inducements.

Summary and conclusions

" I prefer to side with the optimists. The alternative is simply too depressing." [77]

Technical existence proofs seem to exist with respect to combining increasing efficiency of man-made energy systems with renewable energy sources, and carbon-sequestering techno-ecology, into aggregates. These aggregates could be designed to contribute, very substantially, to solving environmental and energy dilemmas.

However, the evolutionary paths needed for this solution to emerge are confronted with very interesting challenges with respect to researching and ameliorating the interplay between perceptions and data. For instance, whereas major progress is being made with respect to photovoltaic systems for capturing solar energy and converting it to energy carriers such as electricity or hydrogen, such systems – although supplying energy with essentially no greenhouse-gas emissions – cannot speed up the rate of sequestering fossil-fuel based carbon dioxide already in the atmosphere. To decelerate the rate of climate change large-scale ecological engineering will be necessary, not only in turning deforestation into reforestation, but also in designing and implementing specific vegetation systems providing additional carbon-sink capabilities, as well as biomass for advanced and efficient use of energy .

Tactical and strategical ecological engineering invokes crucial questions of intangible kinds, including how attitudes towards Nature should be analyzed and entered into the energy equation so as to enable local-specific, culturally acceptable and tailormade designs of multiple-use areal systems (such as agro-forestry).

Answers to such questions related to intangibles have to be worked out from truly interdisciplinary (and not merely technocratic) perspectives.

For modernized renewable systems to play a significant role in safe-failing the energy equation, quality measures such as kW of primary energy generated per hectare – although necessary prerequisites – will have to be much further developed. Relevant measures will have to include down-to-earth indicators of the capacity of renewables for providing sustainable development on the local as well as regional and global scale.

As examples towards making progress along these lines one may consider the following.

Specify biospheric stability domain, and its dynamic (resilient) properties.
This is a question for systems theory as well as for disciplines such as biogeo-physics, paleoclimatology, (economic) history, cultural anthropology. A much more clear view of allowable rates of (average and regional) change in climate pa-rameters is needed (including the magnitude and frequency of extreme events).

As long as biospheric stability properties are uncertain, environmental policies should abandon paradigms such as Nature Benign, or Nature Linear, and use a risk philosophy based on the Nature Non-Linear paradigm.
This is because uncertainty as to the implications of the complexity of natural sys-tems (safeguard or trap door?) cuts both ways. Much more research is needed to develop risk philosophies and assessments able to cope with this fundamentally new policy-choice situation.

Wherever possible, designing plans for responding to surprises (safe-failing) has to be a salient task in implementing large-scale renewable resource systems.
Environmental surprise (for good or for bad) has already clearly demonstrated its effects on environmental science and policy (for instance, the ozone-hole story) and is, by definition, a part of the future.

Usable indicators of disruptive change, as well as of sustainable development, must be designed; work to this end should also address the so-called green GNP-measures.
Flip-flop behaviour of natural systems will have to be better understood, as will the concept of increasing frequencies of extreme events. Indicators of progress with respect to sustainable development will have to be formulated for use on the local, regional and global scale. Such indicators would benefit from adopting non-con-ventional perceptions (such as aimed at tracking energy and resource efficiency, and at recognizing the economic values provided by intact ecological systems).

Detailed data-bases and graphical illustrations are needed for nation-specific (even sub-nation specific) contributions to the anthropogenic part of the greenhouse ef-fect.
This would strongly simplify not only work towards international global climate conventions, but would also help science and policy discover as yet hidden options for achieving emissions reductions through a better interplay (optimizing trade-offs) between various societal activities. In this context, much also remains to be done about establishing adequate methods for equating radiatively active trace sub-stances.[78] In particular, the time horizons chosen (the discount rates relative to greenhouse-gas lifetimes) can strongly affect the ranking of available means for emissions reductions. Detailed data bases would also help bridge the gap between perceptions and data in that the particular responsibilities for different societal sec-tors would be clarified. This, in turn, would pinpoint where underlying tensions between soft and hard sciences and attitudes lie.

Obstacles to the communication of scientific information to public policy making, and to the communication of policy issues to the scientific community, have to be scrutinized and, wherever possible, eliminated.
One important step towards bridging the science-policy communication gap would be to promote openess and transparency with respect to underlying inducements.[79] Further, the quest must continue for an interdisciplinarily acceptable tongue, retain-ing sufficient intradisciplinary stringency without jeopardizing multidisciplinary intelligibility (and vice versa).

Acknowledgements
I thank Dean Abrahamson, Thomas B. Johansson, Evan Mills, Lars Nilsson, Deborah Wilson and Ingela M.B. Wiman, and other colleagues at the Department of Environmental and Energy Systems Studies, for constructive criticism of earlier versions of this paper.

Notes and References

[1] Cf.,e.g., the Brundtland Report statement: "Those looking for success and signs of hope can find many: Infant mortality is falling; human life expextancy is increasing; ....../.....But the same processes that have produced these gains have given rise to trends that the planet can no longer bear." (The World Commission on Environment and Development (1987) *Our Common Future,* Oxford University Press, Oxford.) Earlier energy demand scenarios for the year 2020, published in the early 1980s, have been countered by other energy analysts with statements such as : "Meeting the global demand levels projected in the IIASA [International Istitute of Applied Systems Analysis] and WEC [World Energy Conference] studies would require monumental effects to expand energy supplies..."; cf. Goldemberg J., Johansson T.B., Reddy A. K.N. and Williams R.H. (1987) [*Energy for a Sustainable World,* World Resources Institute, Washington], who also detail some of "hidden costs of conventional energy", such as global insecurity, Middle-East Oil supplies in times of crisis, global climatic change and fossil fuel use, linkages between nuclear weapons proliferation and nuclear power. As for the ongoing debate related to whether or not there is consensus with respect to scientific issues in the work in the Intergovernmental Panel on Climate Change (IPCC) (cf. IPCC Policymakers Summary of the Scientific Assessment of Climate Change, Report Prepared for IPCC by Working Group I, June 1990) see, e.g., Kerr R.A. (1990) New greenhouse report puts down dissenters, *Science* 249, 481-482; and the subsequent responses from R.S. Lindzen, W.A. Nierenberg, and A.R. Solow, in turn responded to by R.A. Kerr (*Science* 249, 1093-1094)].

[2] Cf.,e.g.: "...most 'future' studies postulate smooth trends or equilibrium conditions in interactions between development and environment and then seek to identify likely, possible, or even optimal ways to alter them. But history shows that discontinuities, thresholds, and – more generally – surprises are more the rule than the exception in such interactions, exerting a major influence on their outcome..."; see Toth F.L., Hizsnyik E. and Clark W.C. (1989), *Scenarios of Socioeconomic Development for Studies of Global Environmental Change,* International Institute of Applied Systems Analysis, RR-89-4, June 1989. Also: "The centerpiece of an international agreement to protect the world's climate should be a global budget for cumulative carbon releases between now and 2100. This budget should be based on a policy of risk minimization......"; Krause F., Bach W. and Koomey J. (1990) *Energy Policy in the Greenhouse – from Warming Fate to Warming Limit,* Earthscan Publications Ltd, London.

[3] For instance, whether or not somewhere in the far-distant future there could exist a new global *equilibrium* climate (resulting from an anthropogenic contribution to the greenhouse effect) with overall favourable implications for humankind is *not* a pertinent issue. This is because the path towards such potential global climates, i.e. *the rate of change and the consequences of the rate of change (including changes in the magnitude and frequency of extreme events),* is what matters. Global climate change due to alterations of *atmospheric* properties involves a time-constant of around 100 years (cf., e.g., U.S. National Academy of Sciences (1975), *Understanding Climate Change – a Program for Action;* Washington D.C.; p. 22). One century also corresponds to about one or two generations of human beings, or of trees.

[4] Such as acidification, changing climate mechanisms and patterns, threatened biotopes and species fundamental to man.

[5] For an account of this controversy [involving, among other things the role of oceanic dimethylsulphide-releasing plankton in climate regulation, and the so-called Gaia or geophysiology hypothesis] cf., e.g., Wiman B.L.B. (1991) Implications of environmental complexity for science and policy, *Global Environmental Change* (accepted). Also: Wiman B.L.B., Unsworth M.H., Lindberg S.E., Bergkvist B., Jaenicke R. and Hansson H.-C. (1990) Perspectives on aerosol deposition to natural surfaces: interactions between aerosol residence times, removal processes, the biosphere and global environmental change, *Journal of Aerosol Science* 21, *3*, 313-338.

[6] The term "stability" domain is used here to indicate the capability of a system to withstand stress; if stress exceeds this capacity, with respect to absolute magnitude and/or with respect to the rate at which it is applied onto the system, the system will be forced into a new behaviour. As an example, given certain combinations of sunlight, hydrocarbons and nitrogen oxides, not only will smog be switched on but new atmospheric-chemical pathways will also be initiated. The underlying definitions of "stability" in systems theory are dealt with in, e.g., Wiman B.L.B. and Holst J. (1982) *Ekologisk tolerans [Ecological Tolerance];* the Swedish Committee on Natural Resources and the Environment, Swedish Ministry of Agriculture, Stockholm (in Swedish); cf. also Wiman B.L.B. (1991), op. cit., Ref 5.

[7] Cf, e.g., Intergovernmental Panel on Climate Change: Policymakers Summary of the Formulation of Response Strategies. Report prepared for IPCC by Working Group III, June 1990; WMO/UNEP.

[8] Cf., e.g., Hammond K.R., Mumpower J., Dennis R.L., Fitch S. and Crumpacker W. (1983) Fundamental obstacles to the use of scientific information in public policy making. *Technological Forecasting and Social Change* 24, 287-297.

[9] A comparison between the following two approaches ("hard" versus "soft") to the implications of surprising behaviour of systems is instructive: Holling C.S. (1986) The resilience of terrestrial ecosystems: local surprise and global change, in W.C. Clark and R.E. Munn (eds.) (1983) *Sustainable Development of the Biosphere,* Cambridge University Press, Cambridge, pp. 292-317; and Brooks H. (1986) The typology of surprises in technology, institutions, and development, in W.C. Clark and R.E. Munn (eds.) (1986) *Sustainable Development of the Biosphere,* Cambridge University Press, Cambridge, pp. 324-346.

[10] Cf. Wiman I.M.B. (1990) Expecting the unexpected – some ancient roots to current perceptions of Nature, *Ambio* 19, 62-69. Also: Schwarz M. and Thompson M. (1990) *Divided We Stand. Redefining Politics, Technology and Social Choice,* Harvester Wheatsheaf, New York.

[11] See, e.g., Kerr R.A. (1989) Hansen vs. the world on the greenhouse threat. *Science* 244, 1041-1043.

[12] See, e.g., Stavins R.N. (1989) Harnessing market forces to protect the environment. *Environment* 31, 4-7.

[13] Cf., e.g., Wiman I.M.B. (1990), op. cit., Ref 10; Holling C.S. (1986); op. cit., Ref 9; Schwarz M. and Thompson M. (1990), op. cit., Ref 10.

[14] *Safe-fail* can be conceived as a strategy allowing for soft-landing if unpleasant surprise and failure occur; this in contrast to *fail-safe,* invoking the idea that a system could be designed so as to prevent any failure or surprise to happen. As an example, consider the implications of responding the climate challenge through "engineering the unknown – i.e. Nature – out of the equation"; cf. Holling C.S. and Clark W.C. (1975) Notes towards a science of ecological management, in W.H. van Dobben and R.H. Lowe-McConnell (eds.) (1975) *Unifying Concepts in Ecology,* Dr W. Junk B.V. Publishers, The Hague.

[15] Cf. Toth F.L., Hizsnyik E. and Clark W.C. (1989), op. cit., Ref 2.

[16] The concept of "available futures" does not imply any deterministic statement; rather, if one takes the view that there is sufficient scientific rationale for decelerating the rate of global warming it follows that *within a given time-span* there will exist a maximum global temperature regime to comply with. Compare the statement: "What kind of planet do we want? What kind of planet can we get?"; Clark W.C: (1989) Managing planet Earth, *Scientific American,* September 1989, 19-26.

[17] Whether or not anthropogenically driven global warming is already occurring is, in a sense, of little consequence to the indisputable fact that major scientific, policy, and public concerns and debate have, indeed, changed the old environmental agenda into a fundamentally new one. Therefore, the climate-change era is here in at least a psychological sense. Cf. also the following statement: *"We are certain of the following: there is a natural greenhouse effect which already keeps the Earth warmer that it would otherwise be; emissions resulting from human activities are substantially increasing the atmospheric concentrations of the greenhouse gases .... These increases will enhance the greenhouse effect, resulting on average in an additional warming of the Earth's surface. The main greenhouse gas, water vapour, will increase in response to global warming and further enhance it."* IPCC Policymakers Summary of the Scientific Assessment of Climate Change, Report Prepared for IPCC by Working Group I, June 1990.

[18] World Commission on Environment and Development (1987); op. cit., Ref 1.

[19] For a discussion, see, e.g., Lele S.M. (1989) Sustainable Development – a critical review, submitted to *World Development*. Also: Svedin U. (1987) The challenge of sustainability – the search for a dynamic relationship between ecosystem, social and economic factors, Contribution to the International Workshop on Ecological Sustainability of Regional Development, Vilnius, USSR, June 22-26, 1987.

[20] See, e.g., Intergovernmental Panel on Climate Change (1990) Policymakers Summary of the Scientific Assessment of Climate Change, Report prepared for IPCC by Working Group I, June 1; Krause F., Bach W. and Koomey J. (1990), op. cit., Ref 2; Lashof D. A. and Ahuja D.R. (1990) Relative contributions of greenhouse gas emissions to global warming, *Nature* 344, 529-531; Ramanathan V., Cicerone R.J., Singh H.B. and Kiehl J.T. (1985) Trace gas trends and their potential role in climate change, *Journal of Geophysical Research*, Vol 90, No D3, 5547-5566.

[21] These are mainly the greenhouse gases: water vapour, carbon dioxide, methane, chlorofluorocarbons (CFCs), a number of the substitutes for CFCs, and nitrous oxide. To this should be added intricate greenhouse effects of tropospheric ozone, of carbon monoxide, and of stratospheric ozone. Further, the contribution of aerosols should not escape notice, although their net effect on global albedo reamins an open question; cf., e.g., Wiman *et al.* (1990), op. cit., Ref 5; and Hansen J.E. and Lacis A.A. (1990) Sun and dust versus greenhouse gases: an assessment of their relative roles in global climate change. *Nature* 346, 713-719.

[22] Including due consideration of life-times, warming potential per molecule, or per kg, and other pertinent factors; cf. Lashof D.A. and Ahuja D.R. (1990); op. cit., Ref 20.

[23] Lashof D.A. and Tirpak D.A. (eds.) (1989) *Policy Options for Stabilizing Global Climate*, U.S. Environment Protection Agency, Washington D.C. Also: Trexler M.C., Mintzer I.M. and Moomaw W.R. (1990) Global warming: an assessment of its scientific basis, its likely impacts, and potential response strategies, Background Paper no. 6 to the *Workshop on the Economics of Sustainable Development*, Washington, DC, January 23-26, 1990.

[24] See the Appendix.

[25] See, e.g., Bodlund B., Mills E., Karlsson and Johansson T.B. (1989) The challenge of choices: technology options for the Swedish electricity sector, in T.B. Johansson, B. Bodlund and R.H. Williams (eds.) (1990) *Electricity: Efficient End-Use and New Generation Technologies, and Their Planning Implications*, Lund University Press, Lund, 883-947

[26] Cf. The Federal Environment Agency (1989) *Responsibility Means Doing Without*, (Umweltbundes Amt), Berlin August 1989.

[27] This is not necessarily a value-laden assumption. Currently, much less effort is devoted to analysing the constraints and opportunities for the solar option than is directed towards researching the nuclear options. The aspect of combining various measures to manage the environmental dilemmas is exemplified by, e.g., "Living in the greenhouse", *The Economist*, March 11, 1989; 97-100.

[28] Controlled fusion as an energy source can hardly be expected to contribute to commercial energy supply within the time-frame where response to minimizing risks for climate change is needed, and is therefore excluded from this particular discussion on the energy equation.

[29] For fission to earn a significant and global place in the equation a number of breakthroughs with respect to public, security and environmental acceptability probably must first occur, such as for: inherently-safe reactor technology, proliferation-resistant fuel cycles, fully-viable waste management technology, diversion-resistant institutional demands and criteria. Opinions differ as to whether such breakthroughs will occur, and if so, when (cf., e.g., Haefele W. (1990) Energy from nuclear power, *Scientific American*, September 1990; Williams R.H. and Feiveson H.A. (1990) Diversion-resistance criteria for future nuclear power, *Energy Policy* July/August 1990). It may be observed that global annual nuclear grid connections are rapidly decreasing (from about 31 GW in 1985 to about 12 GW in 1989/90) [IAEA: Nuclear Power Reactors in the World, 1989]. The present geopolitical situation, and the ongoing debate on the future of the Nuclear Non-Proliferation Treaty, in particular highlight the proliferation aspects [cf. "NPT in serious trouble", *Nature* 347, 213-214].

Further, following the 1980 referendum in Sweden on the future role of nuclear power in the Swedish energy system the parliament decided that nuclear power be phased out by the year 2010; the time-profile for the phase-out has, however, not been settled.

These and other observations would make it reasonable to exclude also fission from this particular discussion of the global energy equation. Nation-specific characteristics of the equation may differ, of course.

[30] Which degree of consensus should be the relevant guideline is, of course, in itself an issue subject to debate. Dissenting attitudes must never be overlooked as they can be the first indicators of paradigmatic shifts being on their way; cf., e.g., Wiman B.L.B. (1988) *Att vidmakthålla naturresurserna [Maintaining Natural Resources]*, Allmänna Förlaget and Institute for Futures Studies, Stockholm.

[31] Cf. Weinberg C.J. and Williams R.H. (1990) Energy from the sun. *Scientific American*, September 1990.

[32] Wiman I.M.B. (1990) Expecting the unexpected – some ancient roots to current perceptions of Nature, *Ambio* 19, 62-69.

[33] Wiman B.L.B. (1990) Natur under påverkan: Gaia eller Kaos? [Nature under stress – gaian or chaotic response?] In L.J. Lundgren (ed.) (1990) *Vad tål naturen? [Tolerance limits of Nature]*. Swedish Environment Protection Board, Report 3738, Stockholm; pp. 13-42 (in Swedish; extended summary in English]; see also Wiman B.L.B. (1991), op cit, Ref 5.

[34] See, e.g., Schwarz M. and Thompson M. (1990) *Divided We Stand. Redefining Politics, Technology and Social Choice*. Harvester Wheatsheaf, New York.

[35] Odum E.P. (1962) Relationships between structure and function in the ecosystem. *Japanese Journal of Ecology* 12, 3, 108-119.

[36] Cf. Wiman I.M.B. (1990); op. cit., Ref 10.

[37] Cf. note 5.

[38] Cf. note 6 on the stability-domain concept.

[39] For an account, cf., e.g., Wiman B.L.B. (1991); op. cit., Ref 5.

[40] Steven E. Lindberg, Oak Ridge National Laboratory, U.S.A. (personal communication).

[41] Cf. Jäger et al. (1988) *Developing Policies for Responding to Climatic Change*. WMO/TD-No. 225; WMO/UNEP. Note that Jäger et al. state the that the choice of such targets as a temperature change less than a given value per decade "would be based on observed historic rates of change that did not put stress on the environmment or society". One might add that such a concept will have to be scrutinized from a far-reaching assessment of the Man-Nature relationships because ecological systems will not move as units if climate regimes change sufficiently rapidly. Instead, the set-up of the systems will change with respect to species distribution, food-webs, energy chains, hydrology, etc. – that is, with major alterations with regard to, for instance, renewable resources (such as biomass, hydro-power, wind power). Whether human societies can cope with rapid climate changes is thus not only a question of the vulnerability of societal infrastructures to the merely physical manifestations of climate change (settlements endangered by sea-level rise, etc.) but also a question of whether a changing composition of ecological systems disrupts the linkages between societies and their dependency on those systems. Moreover, globally oriented

targets of the above type could look very differently when they are translated to regional sensitivity to the rate of change.

[42] It is of interest in this context to note how the scientific community chose to address this type of uncertainty-management *already two decades ago* when – as we now know – there were substantially less data and insight with respect to climate change: "We attach great importance to the identification of the appropriate international forums in which there can be a continuing assessment of those activities of man which may have a serious impact globally or in large geographic regions. Through these forums *agreements should be sought for common national policies and programs that will avoid or reduce the impacts which may jeopardize the globe or large regions.*" (Emphasis added) (*Inadvertant Climate Modification. Report of the Study of Man's Impact on Climate (SMIC)*, Sponsored by the Massachusetts Institute of Technology, hosted by the Royal Swedish Academy of Sciences and the Royal Swedish Academy of Engineering Sciences; the MIT Press, Cambridge, Massachusetts, 1971.)

[43] Goldemberg *et al.* (1988); op. cit., Ref 1.

[44] Bodlund B. *et al.* (1989); op. cit., Ref 25. Cf. also Buderi R. (1990) Utilities see the green light, *Nature* 343, 399.

[45] Goldemberg J., Johansson T.B., Reddy A.K.N. and Williams R.H. (1985) Basic needs and much more, with one kilowatt per capita. *Ambio* 14, 4-5, 190-200.

[46] Cf., e.g., Kahane A. (1989) Conference report: Sweden's energy path. *Energy Policy*, December 1989.

[47] Cf., e.g., Goldemberg *et al.* (1988), op. cit., Ref 1; Fickett A.P., Gellings C.W. and Lovins A.B. (1990) Efficient use of electricity, *Scientific American*, September 1990.

[48] Goldemberg *et al.*, (1988); op. cit., Ref 1. See also Bodlund *et al.* (1989); op. cit., Ref 25.

[49] Evan Mills, Department of Environmental and Energy Systems Studies; personal communication; see also Mills E. (1990) Sweden's acid test: planning for economic growth, the nuclear phase-out, and reduced CO_2 emissions. Invited paper for Enhancing Electricity's Value to Society, Canadian Electrical Association, Toronto, Canada, October 22-24, 1990.

[50] Bodlund *et al.* (1989); op. cit., Ref 25.

[51] The term *carbon* implies *carbon equivalents*. This puts, implicitly, a strong constraint on the supply strategies. For instance, energy forests managed through high-level nitrogen fertilization would be geared to nitrous oxide emissions potentially offsetting some part of the gain in CO_2 reductions in the supply mix.

[52] It should be noted that freezing CO_2 emissions can only be a first step for Sweden in meeting the challenge by the Working Group I of the Intergovernmental Panel on Climate Change (IPCC WG I, June 1990, op. cit.); stating that: "the longlived gases would require immediate reductions in emissions from human activities of over 60% to stabilise their concentrations at today's levels; methane would require a 15-20% reduction". (Note, that stabilizing the *concentrations* of greenhouse gases does not imply *climate* stabilization.) Cf. also Kelly M. (1990) Halting global warming. In J. Legget (1990) *Global Warming. The Greenpeace Report*. Oxford University Press, Oxford; pp. 83-112.

[53] Wright D.H. (1990) Human impacts on energy flow through natural ecosystems. *Ambio* 19, 189-194.

[54] See, e.g., Reaid W.V. and Miller K.R. (1989) *Keeping Options Alive*. World Resources Institute, Washington D.C.

[55] E.G., Sedjo R.A: (1989) Forests – a tool to moderate global warming? *Environment* 31, 1, 15-20. Consult also the appendix going into some detail on this point.

[56] Wiman *et al.* (1990); op. cit., Ref 5.

[57] CF. Weinberger C.J. and Williams R.H. (1990); op. cit., Ref 29.

[58] Cf., e.g., Wiman I.M.B. (1990); op. cit., Ref 10.

[59] Cf., e.g., Winterbottom R. and Hazlewood P.T. (1987) Agroforestry and sustainable development: making the connection. *Ambio* 16, 100-110.

[60] The Board of Science and Technology for International Development (BOSTID), linked to the US NAS has recently produced a set of papers on the theme "tools for fixing a broken planet".

[61] Cf. "The vanishing jungle", *The Economist*, October 15, 1988; 25-28.

[62] Cf. Wiman I.M.B. (1990); op. cit., Ref 10.

[63] Hughes J.D. (1975) *Ecology in Ancient Civilizations*. University of New Mexico Press, Albuquerque.

[64] In order to get an idea of the current technical limitations (i.e. regardless of cultural, ecological, economic and other constraints, including climate modifications that could be induced) in harnessing the "prodigious" rate of solar input to the Earth's atmosphere consider the following thought experiment. The area of the Saharan desert is around $5 \cdot 10^8$ hectare. If used for a photovoltaic power plant this area could generate around 100 TW, or about a ten times the current output from anthropogenic energy systems but only a very minor fraction (less than 0.06%) of the solar energy input to the atmosphere.

[65] This comparison (14% versus 75%) is not wholly compatible, however; a major fraction of Sweden's use of domestic energy sources relates to digester liquors and refuse, with only a small fraction emanating from direct input of energy from wood fuels.

[66] Houghton R.A. (1990) The future role of tropical forests in affecting the carbon dioxide concentration of the atmosphere. *Ambio* 19, 204-209.

[67] Firor J. (1988) Public policy and the airborne fraction. *Climatic Change* 12, 103-105. (Guest Editorial) In particular (according to Firor's idea) the "helping hand" offered by the physico-chemical absorption of CO_2 by the oceans should be studied much more closely. One point in Firor's reasoning is that his concept will only work for low excess-carbon concentrations in the atmosphere (the relative oceanic sink rate for carbon diminishes with increasing atmospheric CO_2-concentrations). He therefore concludes that "negative fossil fuel growth scenarios should be carefully studied both by those who model the carbon cycle and those who are concerned with policy analysis."

[68] For instance, as an order-of-magnitude exercise, assuming a 25% efficient solar panel, about 4 to 5 square meters would suffice to provide the domestic, electrical needs (excluding space heating) of an average Swedish household per year.

[69] Cf.,e.g., Akbari H., Huang J., Martien P., Rainer L., Rosenfeld A., and Taha H. (1988) The impact of summer heat islands on cooling energy consumption and global CO_2 concentration. Paper presented at the ACEEE Summer Study on Energy Efficiency in Buildings, Asilomar CA, August 1988.

[70] Krause F., Bach W. and Koomey J. (1990) *Energy Policy in the Greenhouse. From warming Fate to Warming Limit*. Earthscan Publications Ltd, London.

[71] Cf., e.g., Nilsson J. and Grennfelt P. (eds.)(1988) *Critical Loads for Sulphur and Nitrogen*. UN-ECE and the Nordic Council of Ministers.

[72] Cf. Goodman G.T. (1987) Biomass energy in developing countries: problems and challenges. *Ambio* 16, 111-119.

[73] Cf. Holling C.S. (1986); op. cit., Ref 9.

[74] Cf.,e.g., Toth F.L. (1988) Policy exercises. Objectives and design elements. *Simulation and Games* 109, 235-255.

[75] Cf., e.g., Hammond K.R., Mumpower J., Dennis R.L., Fitch S. and Crumpacker W. (1983) Fundamental obstacles to the use of scientific information in public policy making. *Technological Forecasting and Social Change* 24, 287-297.

[76] See, e.g., Hendersson-Sellers A. (1990) Australian public perception of the greenhouse issue. *Climatic Change* 17, 69-96.

[77] Brewer G.D. (1986) Methods for synthesis: policy excercises. In W.C. Clark and R.E. Munn (eds.) (1986) *Sustainable Development of the Biosphere*. Cambridge University Press, Cambridge, pp. 455-475.

[78] Cf., e.g., Smith K.S. and Ahuja D.R. (1990) Toward a greenhouse equivalence index: the total exposure analogy. *Climatic Change* 17, 1-7.

[79] Cf. Ravetz J. (1990) Knowledge in an uncertain world. *New Scientist* 22 September, 1990.

APPENDIX

This appendix serves as a data support document (including details, uncertainties, references and words of caution) for figures given in the main text, and deals with the following items:

Table A1: estimated greenhouse-gas emissions (carbon equivalents) from Sweden;
Table A2: estimated solar energy inputs (kW year per year and hectare);
Table A3: estimated gross data for biological and anthropogenic energy-systems outputs (kW year per year and hectare);
Table A4: estimated primary energy outputs from some renewable-resources design components (kW year per year and hectare);
Table A5: a summary table for A2 through A4 selecting the data chosen for representation in Figure 8 in the main document);
Table A6: examples of potentials for making specific primary-enery use use per unit of given goods/ services) more efficient;
Table A7: examples of carbon-sink capacities for some plants or plant systems.

Please observe footnotes (format: [A#]) to these tables.

Table A1. Estimated current (approx. 1989) emissions of greenhouse gases from Sweden.[A24] Unit: Mtons of carbon equivalents; equating based on Lashof D.A. and Ahuja D.R. (1990) Relative contributions of greenhouse gas emissions to global warming, *Nature* 344, 529-531; Rodhe H. (1990) A comparison of the contribution of various gases to the greenhouse effect, *Science* 248, 1217-1219; Wilson D. (1990) Quantifying and comparing fuel-cycle greenhouse-gas emissions from coal, oil and natural gas consumption, *Energy Policy* July/August 1990. (The time scale 100 years is used in integration of the equation for the accumulated greenhouse effect for each gas.) Note that there is considerable uncertainty associated with the table, partly from the limited data available at present, partly from limitations in atmospheric chemistry and physics models (e.g., for the indirect greenhouse-forcing effects of CO; the amount of O3 formation caused by CH4 and other hydrocarbons; the effect on stratospheric H2O caused by CH4; the decay rate of various gases, in particular CO2.)

Sectors: HC=home & commercial, ID=industry, TP=transportation, CFC=CFC-use, AG=agriculture, FO=forestry, MI=miscellaneous[A1], TBG=total by gas, PBG=percent by gas)

sector> gas	energy use			CFC	AG	FO	MI	TBG	PBG
	HC	ID	TP						
CO2	6	4.2	5.5[A2]	-	0[A3]	[A4]	0.1[A5]	15.8	29.9
CH4	-	-	0.01[A6]	-	3.3[A7]	[A8]	6[A9]	9.3	17.6
N2O	[A10]	[A11]	0.3[A12]	-	[A13]	[A14]	[A15]	0.3	0.6
CFCs	-	-	-	21[A16]	-	-	-	21	39.7
CO	[A17]	[A18]	6.4[A19]	-	[A20]	[A21]	[A22]	6.4	12.1
O3[A23]	?	?	?	?	?	?	?	?	?
Total by sector	6	4.2	12.2	21	3.3	?	6.1	52.8	
% by sector	11.3	7.9	23.1	39.7	6.3		11.5		100

Table A2. Some solar-energy input data; unit is kW year per year and hectare.[A26]

Solar radiation incident on top of the Earth's atmosphere (solar constant): 13530 ($\pm$210).

Energy received at the Earth by a horizontal surface, yearly total, clear conditions with diffuse contribution included [actual values vary due to cloud, aerosol concentrations etc., estimates given within square brackets]

Equator (0°)	Tropics (23.5°)[A25]	Mid-earth (45°)	South Sweden (56°)	Polar circle (66.5°)
2625 [1200]	2625 [1300]	2167 [975]	1940 [870]	1598 [700]

Table A3. Some data for biological and anthropogenic energy systems.

Total net primary productivity of the biosphere	100 TW [A27]
Average biospheric net productivity per unit area ($\approx$0.001 of solar energy received)	2 kW/ha
Total anthropogenic energy output (primary energy use) (whereof c. 1.5 TW from biomass?)	11.8 TW [A28]
Average anthropogenic energy output per unit area (whereof c. 0.026 kW/ha from biomass?[A29])	0.2 kW/ha
Anthropogenic channelling of biospheric net primary productivity (whereof c. 0.7 TW to human energy systems?) (cf. [A27,A28])	18-35 TW
Average channelling of net primary productivity (whereof c. 0.013 kW/ha to human energy systems?)	0.35-0.70 kW/ha

---

Table A4. Estimates of primary energy output from some renewable sources in relation to area demand (kW year per year and hectare).[A45,A46,A47]

Wave	high:	66[A30]	medium:	26[A31]	low:	9[A32]
PV	high:	550[A33]			low:	100[A34]
PV/H2	high:	169[A35]			low:	112[A36]
Heat via solar thermal	high:	700[A37]	medium:	428[A38]	low:	200[A39]
Wind energy	median-winds 7-9 m/s:			7[A40]		
Solar-thermal electric	high:	42[A41]			low:	20
Biomass	high:	27[A42]	medium:	6[A43]	low:	2.5[A44]

Table A5. Summary of solar energy inputs/outputs in kW year per year and hectare. Energy penalties for tapping the various sources are not included (if incluced output values would be reduced by around a factor of 2). See tables A2 to A4 for detailed comments.

Energy source		Representative value for Figure 8
geophysical inputs (kW per hectare)		
solar constant	13530 (±1.5%)	13530
energy received, clear conditions		
equator	2625	
tropics	2625	
mid-earth	2167	2000
south Sweden	1940	
polar circle	1598	
energy received, practically		
equator	1200	
tropics	1300	
mid-earth	975	1000
south Sweden	870	
polar circle	700	
potential energy output from technological solar engineering (kWelectricity per hectare)		
solar panel in space (per cell area)	3856	4000
PV panel, 28%, efficiency, tropics	550	500
PV plant, 10% cell efficiency,per site area	200	200
solar heating, high	230	
medium	143	150
low	66	
PV-hydrogen (10% cell efficiency, c. 20 mm rainfall needed per year)	170	
		150
PV-hydrogen (15% cell efficiency) (c. 40 mm rainfall needed per year)	110	
wave energy, high	66	
medium	26	30
low	9	
solar thermal electric, high	40	
(tower plant) low	20	30
wind energy, medium (7-9 m/s)	7	7
(geothermal, global average	0.6	Not used in Fig. 3)
(ocean currents[A48] (per area transversed)	3000 - 80000	Not used in Fig. 3)
(salinity-gradient energy[A49]	?	Not used in Fig. 3)
potential energy output from biological/ecological solar engineering (kWelectricity per hectare)		
(assuming a heating value of 4.4 MWh per ton, and a heat/electricity conversion factor of 1/2.8)		
biomass, high (e.g., sugarcane, elephant grass)	10	
biomass, medium (e.g., willow, poplar)	2	5
biomass, low (e.g., lake/stream productivity)	0.8	

Table A6. Potential for making primary-energy use more efficient, i.e. [energy use for given goods or services in 1988 minus corresponding energy use in about 2010]/[corresponding energy use in 1988].Reliable estimates for agriculture, forestry and fisheries are lacking. [A58]

household appliances	service sector	industry	transportation	space heating
51[A50] to 30%[A51]	55[A52] to 35%	55[A53] to 15%[A54]	75[A55] to 50%[A56]	≈40%[A57]

Table A7. Supposing carbon fixed by plants were harvested and temporarily taken out of the carbon cycle (i.e. protected against decaying back to atmospheric carbon) how much carbon per hectare could be drawn from the atmosphere, and what would be the time horizon for harvesting for various species or plantations? Note that protection against decaying could be achieved in various ways, some of which could enable a societally useful path of the temporally sequestered carbon through society (thus the biomass need not be encapsuled and buried). "Carbon-emission penalties" due to harvesting etc. not included.[A59,A60]

Species or type	tons of carbon fixed per year and hectar	use for carbon sequestering: time-horizon
Barley (grain)	2	
Barley (straw)	1	
Lucerne (alfalfa)	4	
Corn	5	≈ 1 year
Beet (root)	4	
Beet (baulm)	1	
Potato (bulb)	5	
Potato (baulm)	1	
Reeds (Sweden)	5	
Elephant grass (Queensland)	30	
Willow, poplar (Sweden)	6	≈ 10 years
Giant Leucaena (Hawaii)	10 - 15	
Eucalyptus (USA)	13	≈ 20 years
Eucalyptus (Brazil)	10 - 30	
Tropical plantation (young)	1-3	≈ 5 years
Tropical plantation (medium age)	2-7	≈ 15 years
Tropical plantation (mature)	1-5	≈ 35 years
Strategical forest plantations (USA)	5-6	≈ 20-50 years
Tropical rainforest	10	
Savanna	4	
Temperate forest	6	≈ 0 years[&]
Temperate grassland	3	
Swamp, marsh	15	
Upwelling zones	2	
Algal beds and reefs	12	

[&] no net carbon sink unless area of biome expands/is expanded; destruction of these systems would release the amounts indicated

A1 E.g., leakage from waste deposits.

A2 Approx. 85 % from road traffic, 10 % from air transports, 5 % from sea transports.

A3 Closed carbon cycle assumed.

A4 Swedish forests at present net sink for carbon; approx. 10 Mtons of carbon fixed per year in forests (30%) and soils (70%); approx. 2 Mt of carbon fixed per year in Swedish peat bogs; the methane release from these bogs corresponds to approx 1 Mt (thus net annual carbon fixation in peat approx. 1 Mt). Net annual fixation of carbon in Swedish lake and sea waters uncertain, probably strongly variable between years due to algal blooms, etc.

A5 Estimated net release of non-bio-carbon from household waste combustion.

A6 Estimated from emission factor of 80 mg CH_4 per km. Estimated indirect greenhouse effect of methane included.

A7 Mainly livestock.

A8 Unknown, probably negligible.

A9 Essentially leakage from waste deposits.

A10 Data extremely uncertain, probably not insignificant.

A11 Data extremely uncertain, probably not insignificant.

A12 Essentially due to current use of 2- and 3-way catalyst in transportation; value expected to increase by a factor of five to six by the year 2000.

A13 Data extremely uncertain, not likely to be insignificant.

A14 Data extremely uncertain, not likely to be insignificant.

A15 No data; probably small but not insignificant; for instance, increasing installation of N-reduction facilities for waste-water treatment will increase N_2O-emissions.

A16 Decreasing.

A17 No data; probably insignificant relative to transportation.

A18 No data; probably insignificant relative to transportation.

A19 Indirect effects through interaction with hydroxyl radicals and methane.

A20 No data, probably insignificant.

A21 No data, probably insignificant.

A22 No data, probably insignificant.

A23 Generation of tropospheric ozone; no data. Possibly decreasing from transportation sector, decrease possibly balancing increasing N2O (cf. note 12).

A24 Based on my own calculations and a number of sources, including *Luft '90 (Air '90): Action Program against air pollution and acidification;* Swedish Environment Protection Board, 1990, Solna (in Swedish); Calander K., Gustavsson K. Grennfelt P., Lövblad G. (1988) *Emission av koldioxod från antropogena källor i Sverige (Emission of carbon dioxide from anthropogenic sources in Sweden),* IVL Report L88/365, December 1988 (in Swedish); Swedish Natiomnal Energy Board (1989) *Energi och koldioxid (Energy and carbon dioxide),* PM 1989-02-03 (in Swedish); Perby H. (1990) *Lustgasemission från vägtrafik (Nitrous oxide emissions from road traffic);* Swedish Road and Traffic Institute; VTI Report 629 (in Swedish); Transportforskningsberedningen (Swedish Transportation Research Commission) (1990) *Luftföroreningar från trafik (Air pollution from traffic),* April, 1990 (in Swedish); World Resources Institute (1990) *A guide to the Global Environment, 1990-91;* Levander T. (1989) Swedish National Energy Board, PM 1989-09-14.

A25 Note that the hours of daylight are greater in the tropices than at the equator when the sun's path is almost directly overhead.

A26 Cf., e.g., Sayigh A.A.M. (ed.) (1977) *Solar Energy Engineering,* Academic Press, New York; Brinkworth B.J. (1972) *Solar Energy for Man,* The Compton Press, Salisbury.

A27 Estimates differ, however. Wright D.H. (Human impacts on energy flow through natural eceosystems, and implications for species endangerment; *Ambio* 19, 189-194, 1990] gives the value 77 to 89 TW for the Earth's natural *terrestrial* systems; Grubbs M.J. [The Cinderella options, *Energy Policy* July/August, 1990] gives 112 TW; Diamond J.M. [Human use of world resources, *Nature* 328, 479-480, 1990] gives 133 TW. According to Wright [op. cit.] about 20% to 30% of the NPP (net primary productivity) is being diverted (or prevented) from flowing through natural systems by human activities, i.e. about 18 TW to 27 TW (Wright explicitly does not consider biomass for fuel-use).

Diamonds [op. cit.] arrives at about 27% (of about 133 TW) being used, diverted or permanently reduced by humans, i.e. about 35 TW)

A28 Davis G.R. (1990) Energy for planet Earth, *Scientific American* September 1990. The value given is the primary energy input to the world energy system; the amount finally used for transportation is 1.9, for industry 3.0, and for residential and commercial 3; about 3.8 TW is lost or, to some extent (about 0.78 TW) channelled to non-energy use. About 1.6 TW, or about 13%, of the energy input to the anthropogenic energy system come from biomass (plus minor contributions from solar technology and wind power). Diamond [cf. note 27] estimates that about 0.5%, or c. 0.7 TW, of global net primary productivity is directly diverted to energy use, clearly inconsistent in detail with Davis and others, although within a factor of 2 of current estimates. Hence, note the distinction between renewable energy input to the human energy system (about 1.6 YW), and the fraction of total net primary productivity (around 25 TW or so) used for a variety of purposes by man.

A29 For comparison: Sweden's energy output is c. 0.053 TW, or 1.3 kW/ha; of this, digester liquours, wood and refuse account for c. 14%, or 0.18 kW/ha (direct biomass use c. 0.003 kW/ha); whereas in Africa commercial energy amounts to c. 0.54 TW, or 0.18 kW/ha, to this should be added about 1/(1-0.75)x0.54 TW=2.16 TW from non-commercial biomass energy, i.e. 0.72 kW/ha. That is, roughly 2% of energy use in Sweden, or about 0.0003% of solar energy received, comes from "directly channelled solar energy"; whereas about 75% of energy use in Africa, or about 0.06% of annual solar energy received, comes from such energy sources. Sweden contributes 0.15% to world population, Africa 12.2 %. (Calculations based on Goldemberg J., Johansson T.B., Reddy A.K.N. and Williams R.H. (1988) *Energy for a Sustainable World*, Wiley Eastern LtD, New Delhi; World Resources Institute (1990) *A guide to the Global Environment*, 1990-1991; Miller A.S., Mintzer I.M. and Hoagland S.H. (1986) *Growing Power*, World Resources Institute, Study 5, April 1986; *Energy in Sweden (1989) - Facts and Figures*, Swedish National Energy Board; and other sources.

A30 Type: buoy power plant, linear electricity generators.

A31 Type: oscillating air compression, pressure-driven turbines.

A32 Type: buoy power plant, saline-water pumps plus turbines.

A33 Laboratory point-contact crystalline cell, applied in the tropics; not demonstraed in field application. If positioned in suitable space orbit the value would be around 4000.

A34 Si-cell plant; electricity generated per plant site area.

A35 Solar cell efficiency 15%.

A36 Solar cell efficiency 10%.

A37 Assuming 50% of incident solar radiation(average between 1000 and 1400 kW/ha) can be utilized.

A38 Possibly practical value.

A39 Assuming 20% of incident solar radiation can be utilized.

A40 Power plant (or farm) area not necessarily withdrawn from multiple use.

A41 Tower-construction type.

A42 E.g., sugar cane, elephant grass. Intensive aquaculture of carefully designed algal-species successions, or of water hyacinths, could approach or exceed this value.

A43 E.g., energy-forest plantation, energy-grass plantation. As for the competing aspect of biomass use in addressing the build-up of CO_2 in the atmosphere, namely the carbon sequestering capacity of various forest types, these vary (within a factor of five) around 6 tons of carbon per hectar and year; the real sequestering capacity is of course dependent upon the length of the rotation period and the way the biomass is handled after harvest (cf., e.g., Sedjo R.A. (1989) Forests - a tool to moderate global warming? *Environment* 31, 1, 15-20).

A44 E.g., lake or stream productivity.

A45 Energy penalties due to growing, harvesting, transportation etc. not included.

A46 Estimates based on a number of references, including Falkemo C. (1980) *Vågenergiboken (The Wave-Energy Book)*, Ingenjörsförlaget, Stockholm; Goldemberg J., Johansson T.B., Reddy A.K.N. and Williams R.H. (1988) *Energy for a Sustainable World*, Wiley Eastern Ltd, New Delhi; Grubb M. J. (1990) The Cinderella options: a study of modernized renewable energy technologies - part 1, technical assessment. *Energy Policy* July/August 1990; Ogden J.M. and Williams R.H. (1989) *Solar Hydrogen - Mowing Beyond Fossil Fuels*, World Resources Institute, October 1989; The Swedish Commission for

Energy Research (1979) *Förnybara energikällor [Renewable Energy Sources]*, DFE-Report nr 21; Ericson S.-O. (1988) *Biobränslen [Biofuels]*, The Swedish State Power Board (in Swedish), FUD-Report U(M) 1988/40; Leif Gustavsson, The Government's Environmental Delegation for SW Scania (personal communication); Weinberg C.J. and Williams R.H. (1990) Energy from the sun, *Scientific American* September 1990; Boardman N.K. (1977) The energy budget in solar energy conversion in ecological and agricultural systems, in R. Buvet, M.J. Allen and J.-P. Massué (eds) (1977) *Living Systems as Energy Converters*, North-Holland Publishing Company, Amsterdam; Lyttkens J. and Johansson T.B. (1981) Renewable energy in western Europe - potential and constaints, in G.T. Goodman, L.A. Kristoferson and J.M. Hollander (eds.) (1981) *The European Transition from Oil: Societal Impacts and Constraints on Energy Policy*, Academic Press, London; U.S. Department of Energy (1990) The Potential of Renewable Energy, an Interlaboratory White Paper (SERI/TP-260-3674).

A47 Certain renewable, or semi-renewable sources such as hydropower, peat, geothermal, salinity (osmotic) power, and others, are not discussed here for a variety of reasons. For instance, environmental concerns (including the indirect generation of CO_2/CH_4-emissions from flooded areas for hydro power; the landscape impact and CH_4-emissions from peat-mining) or current lack of technological progress (osmotic power) place these sources in a very different class of renewables as compared with those addressed above.

A48 In this context tidal power is of interest. For a brief account of one such system (The Severn Barrage) see Everest D. (1988) *The Greenhouse Effect. Issues for Policy Makers*. Policy Studies Institute, Royal Institute of International Affairs, London; pp. 34-35.

A49 Physically, a salinity difference opf 3.5 % corresponds to a fall height of c. 240 m. How, if at all, this source could be harnessed, and what surface area would be required for a power station, is unclear. Cf., e.g., *Saltenergi i Sverige (Salinity energy in Sweden]*, The Swedish Commission for Energy Production, NE 1977:4, Stockholm 1977.

A50 E.g., more efficient refrigerators.

A51 E.g., television equipment etc.

A52 E.g., efficient lighting technology.

A53 E.g., lighting technology.

A54 E.g., electrolysis, grinding processes.

A55 Cf., e.g., Volvo prototype LCP 2000 with commercially available cars in 1989.

A56 Cf., e.g., Toyota prototype AXV with commercially available cars in 1989.

A57 For public buildings (Leif Gustavsson, the Governmental Delegation for western Scania, personal communioation).

A58 Estimates based on Bodlund B., Mills E., Karlsson and Johansson T.B. (1989) The challenge of choices: technology options for the Swedish electricity sector, in T.B. Johansson, B. Bodlund and R.H. Williams (eds.) (1990) *Electricity: Efficient End-Use and New Generation Technologies, and Their Planning Implications*, Lund University Press, Lund, 883-947; Goldemberg J., Johansson T.B., Reddy A.K.N. and Williams R.H. (1989) *Energy for a Sustainable World*, World Resources Institute; Mellde R.W., Maasing I.M. and Johansson T.B. (1989) Advanced automobile engines for fuel economy, low emissions, and multifuel capability, *Annu. Rev. Energy* 14, 425-444. Overall potential for increasing efficiency in the global energy system could be better than 50% over the period 1988 to 2020).

A59 Based on a variety of sources including Goldemberg J., Johansson T.B., Reddy A.K.N. and Williams R.H. (1988) *Energy for a Sustainable World*, Wiley Eastern LtD, New Delhi; Houghton R.A. (1990) The future role of tropical forests in affecting the carbon dioxided concentration of the atmosphere, *Ambio* 19, 202-209; U.S. National Academy of Sciences (1979) *Leucaena - Promising Forage and Tree Crop for the Tropics*, Library of Congress Catalog Number 77-80271; Törner L. (1988) *Tillväxt och energiutbyte vid odling av olika energigrödor på jordbruksmark [Growth and energy contents from cultivating various agricultural crops]*. The National Swedish Energy Board, Report EO-88/9.

A60 Note that the table deliberately neglects the obvious clash between this particular use of plants (i.e. carbon sequestering) and alternative important uses (food, biomass-fuel); however, if carefully designed and well-researched agro-forestry systems are implemented, carbon sequestering could form a part of a multiple-use approach. For instance, whereas *Eucalyptus* species do not mix well with farm crops (because *Eucalyptus* form shallow roots competing with adjacent food species) the deep-rooted *Leucaena leucocephala* (a nitrogen-fixing legume, given the seed is inoculated with an appropriate strain of *Rhizobium* before it is sown) could form a part of an integrated, multiple-use system. Cf.,e.g., Goodman G.T. (1987) Biomass energy in developing countries: problems and challenges, *Ambio* **16**, 111-119.

Environment Protection and Economic Development

A. Hussain

Development Economics Research Program, London School of Economics
and Political Science, Houghton Street, London, WC2A 2AE, UK

My purpose is to draw attention to the links between policies
for environment protection and development of low-income
economies.

I take the phenomenon of global warming for granted and
assume that the present pattern of the use of energy and the
configuration of technology is incompatible with the
maintenance of a stable climate and the protection of the
environment. Global warming, as I understand it, is linked
directly with the increased emission of so-called "greenhouse"
gases and certain chemicals such as fluoro-carbons in the
course of economic and other human activities. For my purpose
what is important is that the emission of such gases is
connected with economic growth.

Economic growth is premised on an increase in energy
consumption in the course of economic activities. Further,
increase in per capita income brought about by economic growth
leads to an increased consumption of energy in the course of
daily living. High energy consumption sustains high per
capita income, and high per capita income leads to a life
style involving high energy consumption.

It is also true that historically high levels of income
have also brought a greater concern with the environment.
People in developing countries are less concerned with the
environment than are people in the developed countries. The
concern with the environment has increased in all developed
economies and the environment has become an important
political issue.

There are two aspects to the problem of global warming:
first, the present scale of the problem and, second, how it is
likely to grow over time. There seems to be disagreement
among experts concerning how serious the problem is. However,
even if it is not serious now it is likely to become so over
time. To indicate what may happen in the future, I turn to
the figures which show the relationship between per capita
energy consumption and per capita income:

A number of features stand out in the Table. There is an
unambiguous positive correlation between per capita income and
energy consumption per head. From the point of view of energy
consumption, humans are highly unequal. One individual in a
high-income economy consumes the same amount of energy as 16
individuals in a low-income economy. This point may be
expressed more dramatically by recomputing the population of
countries in various income classes in terms of an energy
equivalence scale, whereby each individual is assigned a
weight equal to the ratio of his/her actual energy consumption
to the world average of 1,205 kg of oil. The distribution of
world population in energy-equivalent terms is shown in Column

Relationship Between Per Capita Energy Consumption and Per Capita Income:

1 Countries	2 Population (% of World Total)	3 Per Capita Energy Consumption*	4 Energy Equivalent Population** (%)
Low-Income <$5000	2,884 (60.90)	322	722.44 (13.3)
China & India	1,904 (40.20)	424	628.13 (13.3)
Middle-Income $500, $6000	1,068 (22.55)	1,086	902.46 (22.9)
Low to Middle-Income $500, $2000	741.7 (15.66)	797	459.93 (9.7)
Upper to Middle-Income $2000 $6000	326.3 (6.89)	1,766	448.43 (9.5)
High-Income >$6000	784.2 (16.56)	5,098	3,110.6 (65.7)
World	4,736	1,285.22	4,736

* kg of oil equivalent;
** energy equivalent population is equal to; category per capita energy consumption/world per capita energy consumption.

4. The figures in brackets in Column 4 give the percentage share of total world energy consumption by the population in the income category shown.

Comparing Columns 2 and 4, the two distributions are mirror images of each other. Whilst low-income economies with 61% of the world population account for around 15% of world energy consumption, high-income economies with 17% of world population account for around 66%. What is of interest here is not inequality as such but its implication for world energy consumption in the future and for a policy of reducing it. Currently, the middle income economies account for about the same percentage of world energy consumption as they do of world population. For the purposes of discussion here, I shall concentrate on the two extremes of the distribution, low-income economies and high-income economies: low-income economies, because they are of central importance for world development, and also because the rate of growth of energy consumption is particularly high in some of these economies; high-income economies, because it is these economies which have to play the major part in reducing world energy consumption.

Broadly speaking, the process of economic growth implies an increase in per capita incomes. Historically, this has been happening over the last 150 years. For a considerable period of time the process of sustained growth in incomes was confined to only a few countries, but over the last 40 years it has become global. Sustained economic growth is no longer a phenomenon restricted to only a few countries. With continuing economic growth an increasing proportion of the world's population in developing economies will graduate from low-income to at least lower-middle income status. The income per head is at present highly skewed. That is, the left tail of the distribution is very fat.

The scale of the shift of population from low-income to lower-middle income status during the next 30 years (1990-2020) is likely to be much larger than it has been over the last 30 years (1960-1990). Since 1960, a number of economies have succeeded in making such a shift. These countries include some economies of the Far East and the oil-exporting economies. An important feature of these economies is that they are all relatively small in terms of population. Looking forward we may ask which group of countries is likely to make the transition from the low-income to the lower-middle income category during the next 30 years.

Extrapolating from trends during the past two decades, the group will include, amongst others, the two most populous countries: China and India, which contain around 40% of the world's population. The rate of growth of per capita income in China has been exceptionally high since 1978. In the 11 years from 1978 to 1988, Chinese national incomes grew at an average annual rate of 9.5%. Allowing for the rate of population growth of around 1.5 to 1.8%, over the 11 years (1978-1988), Chinese per capita income grew at an annual rate of 7.5 to 8%. Because of the stabilization policy of the Chinese government, in 1989 the growth rate of national income fell to around 3.7% (around 2% growth in income per capita). The Chinese government is planning for a growth rate of 6 to 7% in the 1990s. Given the present population trends (around 1.5% increase in the natural rate), this would mean a growth rate of income per capita of between 4.5 to 5.5%. At this rate, the Chinese population will attain lower-middle income status by the turn of the century. India is likely to follow China around 2010-2015. Thus 25 years from now the distribution of the world's population by income per head will look very different from the way it does at present: it will have a much thinner left tail than it does now.

The transition of China and India to lower-middle income status has highly significant implications for total energy consumption. It is not difficult to calculate the extra demand for energy this will entail. However, projecting from the present-day relationship between per capita income and energy consumption may underestimate the extra demand for energy. It would seem that energy consumption per head in today's low-income countries is higher than energy consumption per head at the same levels of income 30 or 40 years ago. There is an important `demonstration effect' at work. Low-income countries try to emulate the energy-intensive technology and life styles of higher-income countries. Let me take the example of refrigerators and other electrical equipment. The possession of these commodities per 1,000 households in present-day China is much higher than in Japan when it had the same level of per capita income as China has today. The same would appear to hold for motor vehicles.

In sum, the implication is that the rise in world energy consumption per head, due to a shift in the distribution of world population from the low-income category to a higher income category, is likely to be much faster over the next 30 years than it has been over the last 30 years. The important point is that this will happen even if energy consumption per capita in each of the income classes mentioned in the table remains constant. Given the present pattern of energy use, there would seem to be an incompatibility between economic growth and the protection of the environment.

The principal cause of current and prospective global warming is generally considered to be heavy reliance on fossil fuels for generating energy, and the widespread use of certain chemicals such as fluorocarbons. Given present technology for the generation of energy, a large scale replacement of fossil fuels by other fuels which do not have an adverse effect on the environment in the near future (say for the next 30 years) seems unlikely. Thus in the near future, the principal objective of policies for the protection of the environment will have to be a reduction in energy use in the world to a level compatible with the preservation of the global environment.

Given the present distribution of energy consumption in the world, most of the needed reduction will have to take place in high-income economies which are responsible for two-thirds of the world's energy consumption. A percentage point in energy use per head in high-income economies has a far greater effect on total world consumption than a percentage point reduction in low-income economies. Apart from this, high-income economies because of their advanced technology and a relative abundance of capital are in a better position to accommodate a reduction in energy consumption than lower-income economies. Moreover, the development of energy-saving technologies in high-income economies would in time lead to the spread of such technologies in developing countries.

How much reduction in energy consumption per head in high-income economies would be needed? Over time, the reduction will be have to be greater than that which is currently needed to maintain a stable environment. One has to take into account the growth in per capita income in the world in general and in the present-day low-income economies in particular. High-income economies will need to cut energy consumption in order to accommodate the increased energy consumption in developing economies. Such an accommodation is necessary if the policies for the protection of the environment are to work.

The growth of low-income economies such as China and India and the factors which accompany growth will add substantially to the discharge of greenhouse gases in the environment. But it would be wrong to deduce from this fact that it is the developing economies which should keep their energy consumption in check for the sake of preservation of the global environment. Developing economies can rightly argue that the world finds itself in its present precarious position because of the profligate consumption of energy in developed countries. Had high-income economies been more far-sighted and prudent, growth in energy consumption would be causing much less harm to the environment.

How might a reduction in average world energy consumption per head be achieved? Energy consumption is not planned by economies, even less by the world economy as a whole. By and large, total energy consumption in the world is an outcome of the decisions of hundreds of millions of households, firms and organizations. It cannot be reduced by the administrative decisions of governments or international agreements. A feasible policy for a reduction in energy consumption per head has to provide an economic incentive for households, firms and organizations to reduce their consumption and to develop technologies with which to economize.

Under the present economic organization, a higher price is the most effective means of reducing the consumption of a commodity. This simple economic maxim applies to energy with the same force that it does to any other commodity. In economic terms, the problem of global warming consists of two elements. First, the current price of energy is much lower than its social cost, if one includes not only the cost of production but also the adverse effects on the environment. In particular, the present prices of fossil fuels are much lower than those which are needed for the maintenance of a stable global environment.

The second factor concerns the nature of the harm which too high a use of fossil fuels inflicts on the global environment. Some pollution is short-lived or reduced in time by the assimilative power of the environment. That is, it is reversible by human action or normal natural processes. On the other hand, there would seem to be increasing evidence that greenhouse gases lead to changes which are irreversible in terms of the time scale relevant for human activities. An important aspect of irreversibility is that it forecloses certain economic options and may in time require an adaptation on a massive scale. For example, a climatic change on a large scale would render some of the densely populated areas of the world unfit for agriculture or even human habitation. The irreversibility of harm means that the cost of high energy consumption is much higher than initially presumed.

A rise in the price of fuels will have two economic effects. It will lead users to economize on energy. The experience of oil crises since 1974 shows that such possibilities do exist and a rise in price does have a substantial effect on energy use. The second effect of a rise in price will be to encourage the development of energy-saving technologies. The development of technology is affected by prices in the sense that expected profitability of energy-saving technologies depends crucially on the price of energy.

Thus far I have concentrated on the good news associated with a rise in energy prices. Let me turn to the bad news. A rise in energy prices will also have adverse economic effects at least in the short and medium term. It will add to inflation. Keeping the real price of energy constant or raising the price in line with the price index will introduce an additional inflationary factor into national economies. If higher energy prices take the form of higher prices charged by producers, then there will be major distributive effects. Such effects are evident in the case of oil prices and not altogether desirable from the point of view of equity. Oil-exporting economies are generally rich and poor countries are generally importers of oil. The distributive effects are, however, not so serious if higher energy prices are due to higher taxes on energy. Revenue from energy taxes could be used to redirect investment into energy-saving technologies and economic activities, and to compensate poorer households.

Despite its inflationary and adverse distributive effect in the long-term, a rise in energy prices would in the short-run have a deflationary impact and lower rates of economic growth. This was demonstrated by the effect of a sharp jump in the oil price in the 1970s. A substantial increase in energy prices relative to other prices in the economy will disrupt existing patterns of economic activities and life styles. Time and extra investment will be needed for economies to adapt to the new price structure. The experience of the 1970s also shows that the speed with which economies adapt to the new price structure will vary greatly. The implication is that a cut in energy consumption through higher energy prices has a heavy economic cost at least in the short run. It involves a trade-off between short term costs and long term benefits. Further, the costs will be immediately felt but the benefits, although evident to specialists, will not be immediately perceptible to the general public.

Policies for the protection of the environment, it is important to appreciate, involve difficult choices. It would seem that until now, governments in developed and developing economies alike have tended to be short-sighted and opted for immediate economic benefits. Oil prices reached a peak in 1979-1980 following the Islamic Revolution in Iran, and decreased more or less steadily until the outbreak of the current political crisis in the Gulf. Governments, rather than raising energy taxes to keep the oil price high, were content to let oil prices paid by users decrease. The decrease in oil prices helped to bring down the inflation rate and stimulate growth rates. The result has been a partial, if not a complete, reversal of efforts towards the saving of energy stimulated by the two sharp increases in oil prices between 1974 and 1980. It seems that it takes a political crisis in the Middle East to make governments and the general public until that low energy prices may not be permanent.

Let me end with two comments: one concerns developing economies and the second concerns the role which Japan might play in bringing about a reduction in energy consumption in the world.

By arguing that the main burden of adjustment has to take place in developed economies, I do not want to imply that nothing should be done in developing economies. The degradation of the environment is already a serious problem in some developing economies and has a direct bearing on economic growth and the well-being of the population. For example, deforestation may have an adverse effect on global climate but it also has adverse effects in developing economies. These include soil erosion and a more frequent occurrences of floods. But the main point is that if one is thinking in global terms then what happens in developed economies carries a much greater weight than what happens in developing economies. Developing economies need energy-saving technologies at least as much as developed economies. But the main thrust towards the development of such technologies has to come from developed economies. Much of the modern technology in use in developing economies is derived from that used in developed economies.

High-income economies do not form a homogeneous political block. Most significant policies are made at the national rather than at the international level. In the near future it would seem unlikely that high-income economies will agree on an effective set of measures concerning environment protection and energy consumption. The main breakthrough will come from particular countries. In this respect, Japan is in a unique position. It is a high-income economy and a country with an excellent record in adapting and developing technologies for commercial use. Japan, unlike many high-income economies is poor in natural resources. It is more than many other high-income economies affected by disruptions of international energy supplies. Further, during the last 20 years, Japan has set an excellent example of the way in which degradation of the environment may be reversed through concerted action. As the second largest economy in the world, Japan carries economic weight and has the technical capability to set an example to the world of how economical use of energy may be combined with continued growth. I see an important global role for Japan.

Abating Global Warming for Fun and Profit

A.B. Lovins

Rocky Mountain Institute, 1739 Snowmass Creek Road,
Snowmass, CO 81654, USA

Introduction

Abating global warming is generally profitable, without counting the avoided cost of adapting, or failing to adapt, to possible climatic change.

The clearest example arises when fossil fuels are displaced by technologies which use less energy more efficiently to do the same task (or better). When, as is generally the case, those technologies cost less than fuel -- when saving fuel is cheaper than burning it -- then CO_2 is avoided not at a cost but at a profit.

A simple illustration -- one of the costlier examples that could be given -- is the increasingly familiar compact fluorescent lamp. Such a 15-watt lamp can give the same light as a 75-watt lamp, but lasts 13 times as long, and hence more than pays for itself by avoiding a dozen replacement lamps and trips up a ladder, making its electricity savings better than free. Over its life, one such lamp can avoid burning enough coal to keep a tonne of CO_2 out of the air, along with ~20 kg of SO_x and various other pollutants. But the lamp saves tens of dollars' more in utility fuel, lamps, and installation labor than it costs. It thus creates tens of dollars' net wealth, and defers hundreds of dollars' utility investment.

(If the lamp displaces oil- instead of coal-fired electricity, it will save more than enough oil to drive a standard U.S. car from Washington to Chicago, or to drive the most efficient prototype car across the United States and back. If it saves nuclear electricity, it avoids making a half-curie of long-lived wastes and ~25 mg, or 0.5 tonne-TNT-equivalent, of plutonium. These benefits, too, will be achieved at negative cost.)

More broadly, demonstrated technologies can

- save most of the fossil fuel now burned, at a cost far below that of the fuel itself, making this abatement cost less than zero;
- change soil from a carbon source to a carbon sink (thereby reducing related emissions of CH_4 and N_2O too) at a net cost around zero or less; and
- displace CFCs (and often their proposed hydrohalocarbon substitutes) at a net cost close to zero -- though this cost is irrelevant, since the substitution is already required by international treaty in order to abate stratospheric ozone depletion.

The cost of the main global-warming abatements therefore ranges, broadly speaking, from strongly negative to roughly zero to irrelevant. Thus the scientific uncertainties about global warming do not matter for policy: we should do es-

The Global Environment
Editors: K. Takeuchi · M. Yoshino © Springer-Verlag Berlin, Heidelberg 1991

sentially the same things whether the problem is real and imminent or not. So why fiddle while coal burns?

For similar reasons, most of the best ways known today to abate global warming are not inimical but vital to global equity, development, prosperity, and security. Moreover, since they are generally profitable, their implementation does not require dirigiste regulatory intervention, but rather can rely on the intelligent application of market forces.

The perhaps surprising information summarized in this paper rests on a very extensive body of technical analysis and practical experience. This is more fully provided in *Least-Cost Climatic Stabilization* (73 pages), by A.B. and L.H. Lovins -- submitted 15 October 1990 to the Mitchell Prize competition, under public embargo until 6 March 1990, and available from Rocky Mountain Institute. The many primary sources cited there contain full technical details. The cited analyses just on how to save two-thirds of U.S. electricity in lighting, motor systems, and appliances, for example, is in excess of 1,500 pages and 3,000 footnotes. This paper omits all such detail and focuses only on broad principles.

Electric efficiency

Most of the best energy-saving technologies are less than a year old -- especially the superefficient lights, motors, appliances, and other end-use devices that save electricity. Saving electricity gives the most climatic leverage, because it takes 3-4 units of fuel (in socialist and developing countries, often 5-6 units) to generate and deliver a single unit of electricity, so saving that unit displaces many units of fuel, mainly coal, at the power plant. Power plants burn a third of the world's fuel and emit a third of the resulting CO_2, as well as a third of the NO_x and two-thirds of the SO_x (both of which also contribute to global warming -- a little directly, and more by degrading forests and other ecosystems that otherwise store carbon). Electricity is also by far the costliest form of energy, so it is the most lucrative kind of energy to save. Saving electricity saves much capital: the U.S. in the mid-1980s spent as much private capital and public subsidy expanding its electric supply, about $60 billion per year, as it invested in all durable-goods manufacturing industries. Moreover, a fourth of the world's development capital goes to electrification, and about five times as much such capital is projected to be needed in the 1990s as is likely to be available. This ~$80 billion annual shortfall may imperil proposed development, and electrification's capital demands mean foregoing investments in clean water, female literacy, and other vital elements of sustainable development. For these reasons, this discussion emphasizes the frequently undervalued opportunities to save electricity.

A recent reassessment by the Electric Power Research Institute, the U.S. electric utilities' think-tank, found a potential, mainly cost-effective, to save 24-44% of U.S. electricity within this decade, not counting a further 9½-15% already in utilities' demand forecasts or program plans. The California Energy Commission has similarly identified a potential to save electricity 2.5%/y faster than projected load growth. Such electrical savings, and analogous non-electrical energy savings, can save enough money to pay for most *non*-energy kinds of global-warming abatement. Most of this electricity-saving potential is untapped: for the non-Communist world during 1973-87, oil intensity fell by 32%, but non-oil intensity by only 1%.

Today's best electricity-saving technologies can save twice as much U.S. electricity as five years ago, but at only a third the real cost. Still more detailed assessments of these new opportunities, based on *measured* cost and performance data, thus reveal that full retrofit of U.S. buildings and equipment with today's most efficient commercially available end-use technologies would deliver unchanged or improved services while saving far more electricity, and at far lower cost, than previously supposed. This makes it possible to abate a large fraction of global warming at negative net cost.

The modern U.S. electric-efficiency potential includes saving:

- half of motor-system (or a fourth of total) electricity through 35 motor, control, drivetrain, and electric-supply improvements collectively paying back, as EPRI agrees, in ~16 months -- a key opportunity in reindustrializing countries like the USSR, whose motors use 61% of its electricity;
- 80-92% of lighting electricity (the lead National Laboratory found ~80-90%) -- or a fourth of total electricity including net space-conditioning effects -- at negative net cost because of reduced maintenance costs;
- a sixth of total electricity through numerous design improvements to household appliances, commercial refrigeration and cooking, and office equipment (whose potential saving is >90% at zero or negative cost);
- two-thirds of water-heating electricity through eight simple improvements (insulation, high-performance showerheads, etc.);
- most of the electricity used for space-heating and -cooling, through both mechanical-equipment retrofits and improved building shells -- including "superwindows" that can now insulate 2-4 times as well as triple glazing but cost about the same;
- three-fourths of all electricity used in typical U.S. houses and commercial buildings at respective retrofit costs of 1.6¢/kW-h and -0.3¢/kW-h;
- about three-fourths of total U.S. electricity at a net cost averaging about 0.6¢/kW-h (**Figure One**) -- several times cheaper than just *operating* a typical coal or nuclear power plant, even if building it cost nothing. Of course, more could be saved at below long-run marginal cost, which is at least tenfold higher, and higher still when externalities are included.

Potential savings appear to be only slightly smaller and costlier in the most efficient countries than in the United States. Detailed studies have found a potential to save half of Swedish electricity at an average cost of 1.3¢/kW-h, half the electricity in Danish buildings at 0.6¢/kW-h or three-fourths at 1.3¢/kW-h, and 80% in West German households with a 2.6-year payback. To be sure, Europeans do (for example) light their offices less intensively, and turn the lights off more, than Americans do, but that does not affect the *percentage* savings

Figure 1. A Preliminary Estimate of the Full Practical Potential for Retrofit Savings of U.S. Electricity at Average Cost ~0.6 cents per kilowatt-hour

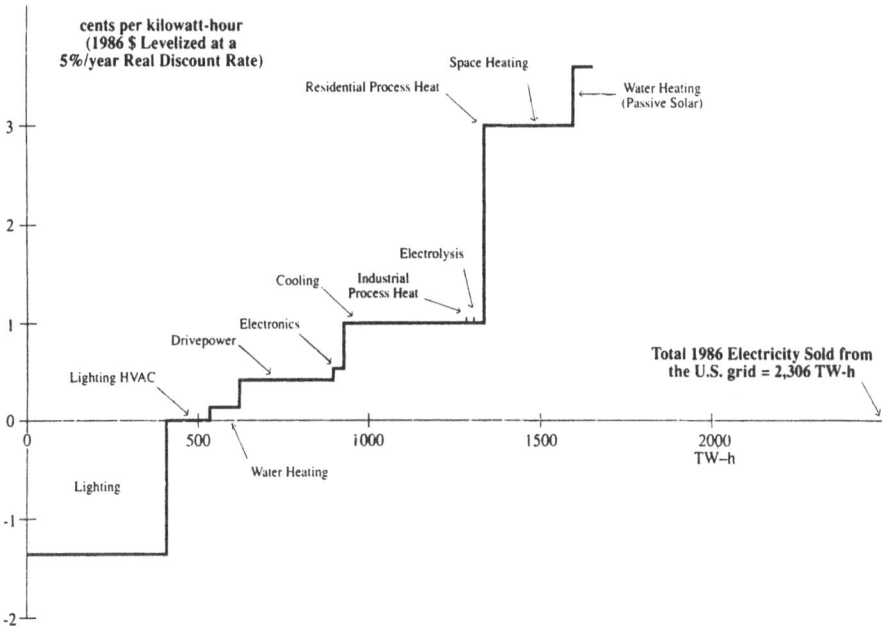

Figure One: Supply curve of the full technical potential to save U.S. electricity by retrofitting the best commercially available 1989 end-use technologies wherever they fit in the 1986 stock of buildings and equipment. The vertical axis is levelized marginal cost, which is negative if the efficient equipment's saved maintenance costs exceed its installed capital costs. The horizontal axis is the cumulative potential saving corrected for interactions. Measured cost and performance data are summarized for about a thousand technologies, condensed into end-use blocks. Fuel-switching, lifestyle changes, load management, further technological progress, and some technical options are excluded. How much of the potential shown is actually captured is a policy variable, but many utilities have in fact captured 70+-90+% of particular efficiency markets in a few months or years through skillful marketing, suggesting that most of the potential shown could actually be captured over a few decades. Note that savings totalling around 50% have a net internal average cost of zero, and that new-construction savings would be larger and much cheaper (often negative-cost) than the retrofit savings shown. For comparison, some utilities report empirical total costs of 0.5¢/kW-h for saving business customers' electricity. The "electronics" and "electrolysis" savings have turned out in more recent analyses to be larger and cheaper than shown, and the drivepower saving, to be probably twice as large as shown, although larger drivepower savings will reduce cooling savings.

available in the lighting energy that *is* used -- a function only of the lighting technology itself, which is quite similar in both places. There is also good reason to believe that Japanese building and drivepower practices offer big savings too.

Abundant observations confirm that the potential savings in socialist and developing countries are much larger and (at world equipment prices) cheaper than in OECD. Differences in what electricity is used for between industrialized and developing or capitalist and socialist countries are surprisingly small, and

217

major savings are available in essentially every application. The feasibility of major electric savings is confirmed by comparisons at all scales: individual technologies, sectoral intensities, and aggregate intensities. It therefore seems reasonable, and probably conservative, to treat the U.S. values as a surrogate for the global average of potential electrical savings' fractional quantity and average cost.

Oil efficiency

The potential for saving oil with today's best demonstrated technologies is also large and cheap. Unlike electricity, about half of the needed technologies are not yet on the market, though they could be within a few years. There are large potential savings in transportation (~44% of world oil use), industrial heat (~12%), building heat (~14%), electric generation (~10%), and feedstocks (~14%). The rest of the oil is used or lost in refineries).

Personal mobility, the most familiar and pervasive use of oil, accounts for about two-thirds of OECD transportation energy use and offers some of the most dramatic savings. For example, ten automakers have demonstrated attractive prototype cars in the range 1.7-3.5 l/100 km (67-138 mi/gal). Some are safer and peppier than today's U.S. cars, and two are said, plausibly, to cost nothing extra to build. Light trucks present similar opportunities to at least triple the efficiency of the world fleet without sacrificing performance. New technologies not yet integrated into a single prototype can probably do considerably better.

Additional major potential arises from a wider mix of vehicle types and sizes, reduced congestion, symmetrical treatment of competing transportation modes, coordinated land-use/transport development, ridesharing, vanpooling, telecommuting, bicycling, walking, bicycling, and public transport innovations (many from Brazil and other developing countries).

Analogous opportunities abound in heavy transport. Boeing's new 777 jet, for example, will be about twice as fuel-efficient as a 727, without using even more efficient unducted fan engines or further innovations. Existing technologies can also at least double the efficiency of heavy trucks, buses, ships, and probably most railways with paybacks of a few years.

Buildings can use simple heat- and hot-water-saving retrofit measures to save upwards of two-thirds of their present fuel use in most OECD countries, more in socialist countries. New options include furnaces up to 97% efficient (while also saving >90% of fan energy), superwindows that gain net winter heat even facing away from the Equator, ventilation heat recovery, and cost-effective ways to insulate or "outsulate" a wide range of existing buildings. Even with 1979 technologies, a major government study found that careful retrofits could save 50% or 75% of U.S. space heat at average costs of $10/bbl and $20/bbl respectively -- severalfold cheaper than heating oil. Technological progress since has cut these costs by probably half, and EPA now considers a 75% reduction in households' total energy intensity achievable by 2025. (Rocky Mountain Institute's headquarters, in outdoor temperatures down to -44°C, has no heating

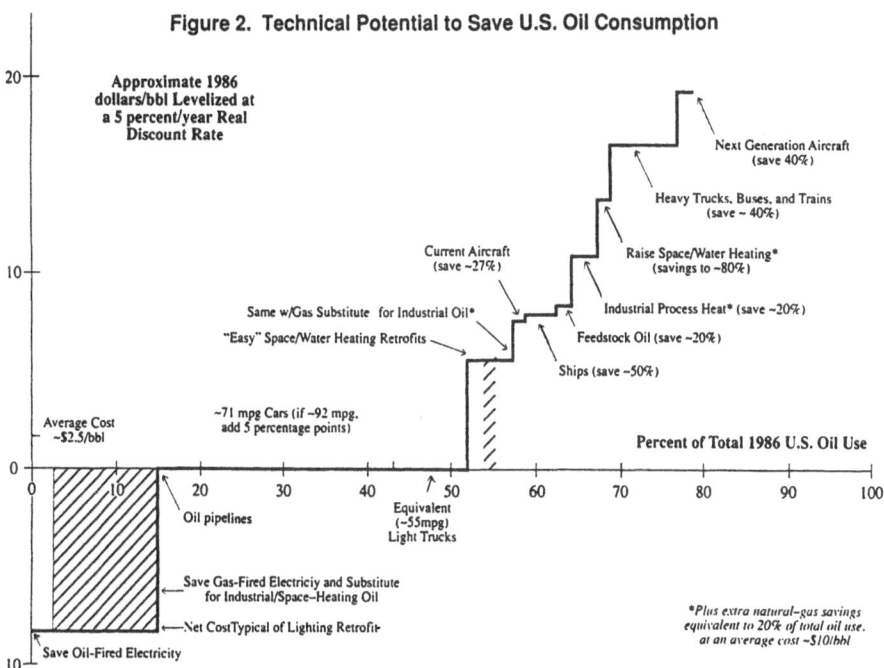

Figure 2. Technical Potential to Save U.S. Oil Consumption

Approximate 1986 dollars/bbl Levelized at a 5 percent/year Real Discount Rate

Next Generation Aircraft (save 40%)

Heavy Trucks, Buses, and Trains (save ~ 40%)

Current Aircraft (save ~27%)

Raise Space/Water Heating* (savings to ~80%)

Same w/Gas Substitute for Industrial Oil*

Industrial Process Heat* (save ~20%)

"Easy" Space/Water Heating Retrofits

Feedstock Oil (save ~20%)

Ships (save ~50%)

Average Cost ~$2.5/bbl

~71 mpg Cars (if ~92 mpg, add 5 percentage points)

Percent of Total 1986 U.S. Oil Use

Oil pipelines

Equivalent (~55mpg) Light Trucks

Save Gas-Fired Electricity and Substitute for Industrial/Space-Heating Oil

*Plus extra natural-gas savings equivalent to 20% of total oil use, at an average cost ~$10/bbl

Net Cost Typical of Lighting Retrofit

Save Oil-Fired Electricity

Figure Two: Supply curve of the full technical potential to save U.S. oil use by retrofitting or substituting the best demonstrated 1988 end-use technologies. The vertical axis is levelized marginal cost. The horizontal axis is the cumulative potential saving (% of total 1986 U.S. end-use) corrected for interactions. Shaded areas represent savings of natural gas that then displaces oil used to heat buildings or industrial processes. The cost and performance data are empirical; costs above $10/bbl are quite uncertain, but this has little effect on the result. No lifestyle changes or intermodal transport shifts are assumed. The curve reflects many conservatisms: *e.g.*, omission of any light-vehicle improvements whose marginal cost exceeds zero, and translating negative-net-cost lighting retrofits (which save the oil- and gas-fired electricity) directly into equivalent $/bbl without taking credit for the value of the fuel displaced. Some of the costlier items, especially aircraft and industrial process heat, are turning out to be cheaper and larger than shown. The overall uncertainty appears to be ~10 percentage points in total quantity and <2x in average cost.

system, uses a tenth the usual amount of household electricity, and repaid its $16/m^2 marginal cost in ten months with 1983 technology.)

Similarly large savings are still available in industrial process heat despite major savings already made. Some large European chemical companies, for example, after halving their fuel intensity since 1973, are now finding *additional* savings *averaging* 70%, with paybacks under two years, just from new methods of thermodynamic process optimization ("pinch technology") coupled with new catalysts. Furthermore, cascading process heat to lower-temperature uses can cost-effectively save ~25-45% of typical OECD usage, and most of the materials produced by industry can be, and are being, rapidly displaced by longer-lived products, lower product mass through better design, computer-aided manufacturing, near-net-shape processing, scrap recovery, remanufacturing, reuse, and recycling.

Adding minor oil savings in feedstocks, the electric industry (via negative-cost electrical savings), refineries, and oil pipelines bring the total U.S. oil-saving potential to ~80% at costs averaging ~$2½/bbl, or about a sixth of the pre-Saddam Hussein world oil price (**Figure Two**). Furthermore, at ~$10/bbl, a further ~20% worth of saved gas would be left over from building and industrial efficiency gains, and could be substituted for vehicular fuels via methanol or compressed natural gas. There is strong evidence that potential fractional oil savings are probably comparable in other OECD countries and generally larger in socialist and developing countries.

Other energy savings

Similarly large and cheap savings are available for other energy forms -- directly used coal, natural gas, and biofuels -- though some of these savings, *e.g.* in Soviet steelmaking and Third World fuelwood, require daunting social reforms that are inseparable from and a prerequisite for those societies' development.

Besides these savings in end-use energy, major savings are available world-wide in energy distribution and conversion. Even in relatively efficient countries like West Germany, for example, recent research by Dr. Florentin Krause at Lawrence Berkeley Laboratory shows that just optimizing the present electrical generating system, *e.g.* via advannced gas-fired cogeneration, could save approximately two-thirds of present utility CO_2 output at negative cost.

Still further options are presented by an immense array of increasingly competitive renewable sources -- many of which are indeed competitive today, most when credited with avoided externalities. A recent U.S. government report, for example, found that with either a small externality credit or accelerated R&D, renewables equivalent to half of today's energy use should be both competitive and installable by 2030 -- plenty to run the country if present cost-effective levels of end-use efficiency are meanwhile adopted. Similar or more sanguine conclusions would apply, too, in most other countries, especially now that advances in high-selectivity surfaces are overcoming cloudiness constraints.

Combined energy opportunities

Together, efficiency and renewables are a powerful prescription for cutting both CO_2 emissions and energy costs. For example:
- A detailed 1989 analysis by the Swedish State Power Board found that doubled electric end-use efficiency (78% cheaper than marginal supply), plus fuel-switching to natural gas and wood, plus environmental dispatch (operating most the plants that emit the least carbon), could together support 54% growth in Swedish real GNP during 1987-2010 and handle the voter-mandated phaseout by 2010 of the nuclear half of the country's

electric generation, yet at the same time *reduce* the heat and power sector's CO_2 output by one-third and *reduce* the cost of electrical services by nearly \$1 billion per year. This result is especially striking because Sweden is arguably the world's most energy-efficient country to start with, with a heavily industrialized economy and a severe climate. Any other country should therefore be able to do better.

- At the same time, a study for the Indian state of Karnataka analyzed the combination of several end-use efficiency measures (but far from a comprehensive list), small hydro dams, bagasse cogeneration, biogas/producer gas, a small amount of natural gas, and solar water heaters. This combination would achieve far greater and earlier development progress than the utility committee's plan, since rejected, yet would use three-fifths less electricity to do so. It would also cost a third as much and emit only 1/200th as much fossil-fuel CO_2. This is encouraging too, since India already emits ~5% of global carbon and projects this fraction, assuming the traditional coal-based strategy, to increase enormously.

These two analyses are especially interesting when considered together, because between them they scope essentially the full global range of energy intensity and efficiency, technology, climate, wealth, income distribution disparities, and social conditions. Yet both find that the money saved by efficiency more than pays for the renewables, yielding a net profit on the whole carbon-displacement package in the energy sector.

Similarly, the Enquête-Kommission of the German Bundestag recently concluded that 80% long-term CO_2 reductions of 80% are both necessary and feasible in OECD; and reports commissioned by the governments of Australia and Canada, and privately for California, similarly found that CO_2 emission reductions of ~20% via energy efficiency would be highly profitable.

Implementing energy efficiency

Of course, no amount of technical potential will reduce CO_2 emissions without implementation. Happily, though, new ways to finance and deliver efficiency (and renewables) to customers, in a wide range of societies, are evolving just as quickly and successfully as the technologies themselves. This paper treats only the more interesting innovations relevant to OECD countries, but analogous ones are becoming available in socialist and developing countries.

Over half of Americans can already get financing from electric utilities for energy-saving equipment, in the form of concessionary loans, gifts, rebates, or leases. Such financing enables the customer, in effect, to use the same low discount rate that the utility otherwise uses for supply-side investments -- rather than one about tenfold higher, as unaided customers do, thus diluting price signals by tenfold. Most U.S. utilities and some governments also support customers with information in a variety of popular and technical forms. Similar information and financing efforts are spreading in Europe and elsewhere in OECD.

Rather than merely marketing "negawatts," many U.S. electric utilities are also starting to make markets *in* negawatts: to make saved electricity into a fungible commodity subject to competitive bidding, arbitrage, derivative instruments, secondary markets, etc. For example, some utilities are:

- buying back savings from customers by paying "generic rebates" per kW-h or peak kW saved -- including rebates for beating government standards or for scrapping old equipment;
- in eight states, operating "all-source bidding" in which all ways to make or save electricity compete at open auction and the utility takes the low bids, which are generally for efficiency (though this approach also enabled the State of Maine to increase its share of private, mainly renewable power generation from 2% to 30% in the seven years 1984-91);
- starting to buy saved electricity from other utilities -- a form of arbitrage on the difference between the cost of supply and efficiency;
- considering making spot, futures, and options markets in saved electricity;
- exploring ways to broker saved electricity between customers, rewarding any customers who goes "bounty-hunting" by correcting ineffiencies anywhere in the system;
- selling electric efficiency in *other* utilities' territories (Puget Power Company, for example, sells electricity in one state and efficiency in nine states); and
- in seven states, experimenting with sliding-scale hookup fees for new buildings -- "feebates" whereby the builder either pays a fee or gets a rebate when the building is connected to the grid (which one, and how big, depends on the building's efficiency). Feebates can offer major economic advantages to all the parties, can generate tens of thousands of dollars' net wealth per U.S. house so built, and are readily coupled with efficiency labelling.

In addition, gas utilities can make money selling *electric* efficiency, thereby changing the behavior of buildings in ways that also help them open up new gas markets. Electric utilities can also sell gas efficiency, and both should be rewarded for selling either. Wisconsin's utility regulators have even ordered that state's utilities to help customers switch to any competing energy form that costs less.

Rapid experimentation in these and other market-making methods has been facilitated by the great diversity of the U.S. electric utility system: ~3,500 utilities of all shapes and sizes in ~50 major and hundreds of minor regulatory jurisdictions. Results so far are encouraging. A few years ago, some utilities had captured ~70-90+% of particular efficiency micromarkets, mainly difficult ones (residential shell retrofits), in only one or two years. In 1990, greater marketing experience has enabled, for example, New England Electric System to capture 90% of a small-commercial pilot market in two months, and Pacific Gas & Electric to capture 25% of its entire new-commercial-construction market in three months.

Such entrepreneurship is being encouraged by a nationwide agreement in principle among U.S. utility regulators to change the rules of price formation so as to ensure that utilities' cheapest options are also their most profitable ones. Although many ways to do this are available, the most common is to decouple utilities' profits from their sales, and then let them keep as extra profit part of what they save their customers. Under a new policy approved in summer 1990, for example, Pacific Gas and Electric Company will be allowed to keep 15% of certain savings -- adding ~$40-50 million to its 1991 profits -- but the customers are better off getting 85% of an actual, prompt saving than getting all of nothing. At this writing, five states have approved such reforms and another 23 are doing so. The previous regulatory scheme rewarded utilities for selling more electricity (or gas) and penalized them for selling less. Similar perverse incentives still exist in many countries, despite the diversity of utility structures, and can be corrected by similar means. Furthermore, a dozen of the United States already take account of externalities in least-cost planning (which, to varying degrees, some 43 states require), and another 17 are currently making rules to do so.

These regulatory moves toward simulating efficient market outcomes have accelerated the already rapid shift in U.S. utilities' culture and mission, away from selling more kilowatt-hours and toward the profitable production of customer satisfaction. About a third of U.S. utilities have already made this transition. Selling more efficiency may reduce their electric sales and revenues, but their costs go down more; and under the new rules, they can keep part of the difference, making money on margin instead of volume. Such a utility can indeed makes money in six ways: it saves operating, construction, and replacement costs, plus associated risks and externalities; under the new U.S. Clean Air Act its fuel savings will be able to generate tradeable emission rights (currently for acid gas, but extendible by future legislation to fossil carbon); as soon as its regulators reform their ratemaking rules, it will be specifically rewarded for efficient behavior; and it can earn a spread on financing customers' efficiency improvements. (Arbitraguers get rich on spreads of a fraction of a percentage point, but the difference in discount rate between utilities and their customers is more like a thousand percent.)

Analogous concepts are starting to enter the oil-efficiency market. Most importantly, on 30 August 1990, the "Drive+" feebate proposal passed the California Legislature by an extraordinary 7:1 margin. (It was then vetoed, but the Governor-elect favors it, so it should pass in 1991.) This bill would enact a revenue-neutral, open-ended sales-tax adjustment based on both fuel efficiency (measured as CO_2 emissions per mile) and smog-forming emissions. Buyers of dirty, inefficient cars would pay two fees; buyers of clean, efficient cars would get two rebates. By influencing car choice directly, feebates overcome the ~4:1 dilution of gasoline prices by the other costs of owning and running a car. Drive+'s self-financing -- the fees pay for the rebates -- attracted broad bipartisan support. Similar proposals are pending in Connecticut, Iowa, and Massachusetts, are being drafted in other states, and have been proposed nationally. Interestingly, General Motors did not oppose Drive+, and at least two makers of superefficient cars, even before its passage, had become seriously interested in rapidly entering the U.S. market.

In short, Drive+ is likely to launch a powerful national and perhaps international trend. A useful early refinement, too, would be "accelerated scrappage": basing rebates for efficient new cars on the *difference* in efficiency between the new car and the old one that is scrapped (if it is worse than a certain level; drivers who scrapped a functioning car and didn't replace it would get a bounty on presenting a death certificate that it had been duly recycled). By offering a far higher price than dealers do for tradeins, the state would put a premium on getting the least efficient, most polluting cars off the road soonest. This incentive would greatly accelerate the energy and pollution savings. It would also enable poor people, to whom the worst cars tend to trickle down, to afford to buy a highly efficient new car that they could then afford to run.

Such feebates have wide application. They could spread from cars, light trucks, and buildings to appliances, aircraft, heavy road vehicles, etc. In each case, they would transfer wealth from those whose inefficient choices impose large external costs on society (global warming, acid rain, oil-import dependence, etc.) to those who save such costs.

Other innovations are also needed for transport efficiency, such as eliminating tax breaks for company cars (which account for the majority of urban commuting in many OECD countries), and the new Stockholm scheme whereby downtown residents who wish to drive their cars during a given month must buy a permit -- which also serves as their free pass to the regional public transit system for that month.

Another useful concept: "golden carrot" rebates designed to elicit the production of specific energy-saving products that are cost-effective but are not yet brought to market because cautious or undercapitalized manufacturers are unwilling to risk retooling costs for uncertain sales. For example, utilities from San Diego to Vancouver may soon join together to pay, say, a $300 rebate for each of the first 10,000 or 100,000 refrigerators sold in their territories that beat the 1993 Federal efficiency standard by at least 50%. Such incremental efficiency would require one or more of the advanced insulating materials mentioned above to be put into mass production. If the refrigerators are not sold, no rebate is paid, so the utility is at no risk of not getting the desired savings; and once that many *are* sold, the manufacturing hurdle has been leapt, the rebates can be discontinued, and continuing sales will then yield far larger savings. A larger "platinum carrot" can then be offered for the next incremental advance, and so on until cost-effective opportunities are exhausted. This proposal was originally developed for North American electric utilities and appliances -- and is now being considered by European and South Pacific utilities too -- but there is no obvious reason why it could not be offered by governments and applied to other products such as light vehicles, in addition to or in lieu of feebates.

The frustrating, though gradually resolving, problems of institutional change must not obscure the major gains already made. During 1979-86, for example, the United States got more than seven times as much new energy from savings as from all net expansions of supply, and more new supply from sun, wind, water, and wood than from oil, gas, coal, and uranium. By 1986, U.S. CO_2 emissions were one-third lower than they would have been at 1973 efficiency levels; the average new car alone expelled almost a ton less carbon per year; annual energy bills were ~$150 billion lower; and annual oil-and-gas savings were three-fifths as large as OPEC's capacity.

During 1977-85, the United States increased its oil productivity at an average rate of 4.8%/y -- four-fifths faster than it had to in order to match both economic growth and declining domestic oil output. (The extra savings halved oil imports.) By 1986, the annual savings, chiefly in oil and gas, were providing two-fifths more energy than the entire domestic oil industry, which had taken a century to build. Oil, however, has dwindling reserves, rising costs, and falling output, whereas efficiency has expanding reserves, falling costs, and rising output.

Capitalizing on such success does require unremitting attention. Had the U.S. simply kept on saving oil as quickly after 1985 as it did during 1977-85, then in no year after 1985 would it have needed any oil from the Persian Gulf. Even today, spending on oil efficiency just the $20-odd billion being devoted to U.S. military costs in the Gulf from August through December 1990 would displace all oil now imported from the Gulf.

The universality of these opportunities is illustrated by a curious and encouraging fact: today, it is the countries, like Japan and West Germany, with the highest energy efficiencies -- those that, not coincidentally, have proven among the toughest economic competitors -- that are now starting to redouble their efficiency efforts as they discover newer and bigger technological opportunities.

The task now for OECD is twofold: to accelerate these historic efficiency gains by harnessing today's far more powerful and cost-effective technologies, *e.g.*, by promoting superefficient cars through feebates with accelerated scrappage; and to extend to electricity the rapid, consistent efficiency gains obtained during 1973-86 for direct fuels. In principle, it should be possible to save electricity as least as quickly as oil, and the best utility programs confirm this: *e.g.*, Southern California Edison's 1983-85 program reduced the decade-ahead forecast of peak load by the equivalent of 8½% of the then-current peak load *per year*, at average program costs of a few tenths of a cent per kW-h saved (~1% of the long-run marginal cost of supply).

Other global-warming abatements

A large empirical literature, much of it assembled in recent studies by and for USEPA and the U.S. National Academy of Sciences, has found that the net cost both of CFC substitution and of reforestation (or avoiding deforestation) is on the order of zero. In addition, the Academy has found that sustainable farming practices, which change farmland from a carbon source to a carbon sink, are at least as profitable as soil-mining and usually more so. Adding these rather large terms to the global-warming abatement potential, therefore, does not significantly increase the average cost of abatement, and may decrease it. Two official studies being published in the U.S. in late 1990, indeed, have found a potential to abate the U.S. contribution by ~50-70% at zero net cost. Adding further opportunities (chiefly in energy efficiency) that those studies omitted or understated increases the abatement potential to a level probably sufficient to stabilize global climate, while reducing its average cost from zero to negative. This result, too, is probably conservative, because it takes no credit for the surprisingly many instances when a single expenditure can abate the emission of more than one greenhouse gas -- a "joint product" overlooked by normal reductionist analyses.

Conclusions

The foregoing summary, together with its supporting material, rebuts eleven prevalent myths (in italics) about abating global warming:

- *Greater scientific certainty should precede action.* The uncertainties about global warming and its potential consequences are substantial, interesting, and likely to cut both ways. But they are also irrelevant to policy, because virtually all the actions needed to abate global warming (if it does turn out to be a real problem) should be taken anyway to save money. These "no-regrets" actions are about enough to solve the problem if it does exist, and are highly advantageous even if it doesn't. The problem with global warming isn't decision-making under uncertainty; it's realizing that in this instance, uncertainty doesn't matter.

- *The issue is whether to buy a "climatic insurance policy" analogous to fire insurance or to defense expenditures (a major investment mobilizing most of the country's scientific and technological resources, and meant to forestall or respond to unlikely but potentially catastrophic threats to national security).* The "insurance" analogy is partly valid, because delaying action until obvious climatic changes are unambiguously underway makes abatement too little, too late, and too costly -- just like trying to install a sprinkler system in a hotel that's currently on fire, or build military forces while you're already under attack, or buy collision insurance after you've crashed your car. Abating global warming will require significant efforts affecting large stocks of people and capital over long periods and with long lead times, so waiting too long will certainly raise cost, difficulty, and risk of failure. But the analogy breaks down, because the real choice is not balancing uncertain future benefits against daunting present costs, but rather making the investment as wisely and quickly as possible in order to achieve both the uncertain future benefits *and* the certain financial *savings*. Any insurance "premium" is actually negative.

- *Abating global warming would be costly.* Distinguished econometricians have claimed that just achieving the Toronto interim target of cutting CO_2 emissions by 20% -- roughly a third of the reduction probably required for climatic stabilization -- would cost the United States alone on the order of $200 billion per year. The econometricians have the amount about right but the sign wrong: using modern energy-efficient techniques to achieve the Toronto target would not cost but *save* the U.S. on the order of $200 billion a year. Their high-cost conclusion is a bald *assumption* masquerading as a fact: the econometric analysis merely asks how high energy prices would need to be, based on historic price elasticities of demand (typically from decades ago), to reduce fossil-fuel use by $x\%$, then counts those higher prices (or their equilibrium econometric effects) as the cost of abatement. This approach ignores the compelling empirical evidence that saving most of the fuel now used is cheaper than even its short-run marginal cost, and hence is profitable rather than costly. Modern efficiency techniques did not exist at the time of the behavior described by the historic price elasticities -- which summarize how people used to behave under conditions that no longer hold and that energy policy aims to change as much as possible.

- *Abating global warming would drastically curtail American and similar lifestyles, and would mean less comfort, mobility, etc.* Nothing could be further from the truth. The fuel-saving technologies that can stabilize global climate while saving money actually provide unchanged services: showers as hot and tingly as now, beer as cold, rooms as brightly lit, torque as strong and reliable, homes as cozy in the winter and cool in the summer, cars as peppy, safe, and comfortable, etc. The quality of these and other services can often be not just sustained but substantially improved by substituting superior engineering for brute force, brains for therm: *e.g.*, with efficient lighting equipment, you get the same amount of light, but it looks better and you see better. The same is broadly true of sustainable agriculture and silviculture, which provide comparable yields with superior quality, resilience, human health, and (generally) profitability.

- *If such cost-effective abatements were available, they would already have been bought.* This is reminiscent, says physicist Murray Gell-Mann, of the econometrician who, asked by his mannerly granddaughter whether she could pick up a $20 bill she'd just noticed lying on the sidewalk, replied, "No, my dear, don't bother: if it were real, someone would have picked it up already." The striking disequilibrium between how much energy efficiency is now available and worth buying and how much has already been bought arises from distinctive, well-understood market failures that leave cheap efficiency seriously underbought at present prices. (For example, consumers have poor access to information and to mature mechanisms for conveniently delivering integrated packages of modern technologies. Discount rates are about tenfold higher for buying efficiency than supply, thus diluting price signals. Many energy utilities misunderstand their business and want to increase their sales -- even though reducing their sales would increase their profits by decreasing their costs even more. Perverse regulatory signals often reward inefficient and penalize efficient behavior. Markets in saved energy are sparse or absent. And present market signals, omitting externalities that may be as big as the apparent fuel prices, make consumers indifferent to whether they buy, for instance, a 20- or a 60-mile-per-gallon car, since both cost about the same per mile to own and drive.) Solutions exist for each of these market failures. These solutions have been proven in market economies and are rapidly emerging in a wide range of other societies, so there is an ample range of effective policy instruments to choose from. The technical and implementation options -- the everyday work of energy-efficiency practitioners -- are mostly unknown, however, to those econometricians who lie awake nights worrying about whether what works in practice can possibly work in theory.

■ *Abatements would be so costly and disagreeable that they could only be achieved by draconian, authoritarian government mandates incompatible with democracy.* On the contrary, the abatements described above are so profitable and attractive that they can be largely if not wholly achieved by existing institutions, within the present framework of free choice and free enterprise. Planners unaware of market-driven alternatives seem anxious to set up new bureaucracies to tell people how to live. Many bizarre schemes have been suggested for substituting dirigisme for markets, penury for development, risks for rewards, and costs for profits. This paper seeks an antidote to that perversion of economic rationality.

■ *Combating global warming requires tough tradeoffs -- swapping one kind of pollution or risk for another.* Abating global warming by resource efficiency can simultaneously reduce or eliminate many other hazards too -- oil-security risks, nuclear proliferation, utilities' planning and financial risks, declining farm and forest yields, etc. -- without creating new ones.

■ *Available means of abatement, singly or combined, will be too small and too slow, so global warming is inevitable and we must start trying to adapt to it.* This counsel of despair is misguided. To be sure, some significant degree of climatic change or increased climatic volatility in some places may already be unavoidable if the more sensitive models prove valid, or if greater climatological or ecological understanding continues to bring unpleasant surprises. A modest degree of adaptation may therefore be prudent if not inevitable: *e.g.*, planning coastal developments to accommodate some sea-level rise and water projects to tolerate shifts in rainfall, or reversing the narrowing of crops' and forests' genetic bases. Nonetheless, the techniques described here, if their benefits are properly understood, show promise of such rapid and widespread deployment that most of the harm projected in today's best models could almost certainly be avoided. Many abatement measures also have the valuable side-effect of increasing resilience in the face of whatever climatic change may nonetheless occur: efficient buildings, for example, are more likely to maintain comfort in weather they weren't designed for.

- *We'll need all the abatement we can get, so let's buy every option that's offered.* This ignores the well-known concept of opportunity cost -- that you can't spend the same money on two different things at once, so each investment foregoes others. If, for example, you spend a dollar on a costly source of electricity to replace coal-burning, such as nuclear power or photovoltaics, it will deliver little electricity per dollar (that's what being expensive means), and so can replace little coal per dollar. If you instead spend the dollar on a cheap source, such as efficiency, it will displace lots of coal. Each dollar spent on a costly option instead of a cheap option, therefore, unnecessarily results in burning extra coal and releasing extra CO_2, simply through not choosing the best buys first. That is why nuclear power makes global warming worse. (Interestingly, the Enquête-Kommission found that 30-45% reductions in West German CO_2 emissions during 1987-2005 were equally feasible whether nuclear power was expanded or phased out.)
- *Abating global warming would lock developing countries into abject poverty, or at least prevent their achieving their legitimate aspirations -- even though most global warming so far has been caused by the industrialized countries.* On the contrary, the abatement options discussed above are not merely compatible with but essential to affordable and sustainable global development and increased equity.
- *Policymakers already know what their options are and haven't chosen those described here, so either the policymakers are stupid or the options don't work.* Many policymakers suppose that abatement must be slow, small, costly, inconvenient, and nasty -- not because that's true, but simply because they don't know any better. The difficulty may rather be the one economist Ken Boulding described: that a hierarchy is "an ordered arrangement of wastebaskets designed to prevent information from reaching the executive." The options described above are available, demonstrated, and often in widespread and successful use. Yet many are so new that they are not yet widely known even to technical experts, and will take many years to filter up to decisionmakers through normal channels. What is needed, therefore, is better and faster technology transfer to the policymakers, so that they begin to think of future energy needs *not as fate but as choice*, to be responsibly and flexibly exercised.

Knowledge and Action for Preserving the Global Environment

P.M. van der Staal

National Institute for Science and Technology Policy,
Science and Technology Agency, 1-11-39 Nagato-cho,
Chiyoda-ku, Tokyo 100, Japan

Now that the threat of nuclear self destruction seems to have diminished as a result of the relaxation of tensions between the superpowers, another Damocles´ sword appears to be hanging over the world. This time mankind seems to be challenging continued human existence more indirectly by changing the global environment on which man depends for survival. Awareness is growing that immediate actions are necessary to prevent future global environmental catastrophes. In order to define the nature of the actions to be taken, the need for more scientific research is often stressed. However, there is no reason to wait for the results of such research before beginning to take action. And what may be even more important than scientific results, is the way in which this knowledge is communicated and used in the decision-making process. This chapter discusses the incorporation of environmental research into the decision-making process.

Introduction

A main reason for organizing conferences on global environment is the growing concern that in the coming century continuous changes in the global environment may cause serious problems for the world population and may even threaten living conditions on this planet. It is assumed that these problems are at least partly caused by human producer and consumer behavior and that concerted actions to control this behavior may help to prevent avoidable and to restore reversible changes in the global environment. In order to understand the problem itself and to determine necessary countermeasures, the need for more scientific research is often advanced, and not only by the scientific community. This research should provide information on present and future changes in the global environment, their natural and human behavioral causes and their impact and consequences for men and societies. Technological research should produce technologies for controlling and reversing these causes and effects. This information should be made available to decision makers in government and industry, and to the general public. The effects of resulting decisions, measures and actions should later be monitored and evaluated by scientific research.
   This scheme, plausible as it seems, assumes traditional roles for science and policy. It is also based on a high esteem for the capabilities and competencies of the participants involved. It finally rests on the application of

The Global Environment
Editors: K. Takeuchi · M. Yoshino   © Springer-Verlag Berlin, Heidelberg 1991

a linear knowledge-action model of description, explanation, prediction and control. Any weakness in these premises will seriously hamper our efforts to handle environmental problems and will lead to long, unproductive and frustrating discussions between people who are seriously concerned with the global environment. For that reason scientists should question these assumptions.

Science can not pretend ever to provide all decisive knowledge in this problem field. Nor need decision makers wait for the availability of this knowledge. The problem is not even this knowledge as such, but its use, dissemination and the way it is translated into the decision-making process. This depends in my opinion on a close relationship between researchers, decision makers and the responsible public.

In the worst case scientists remain in their ivory towers and send their warnings and information, mixed with noise and contradictions, to decision makers. Decision makers in this case quietly decide to ignore scientific advice and blame the professionals, who do not have official responsibilities, for not understanding the political issues involved. Meanwhile the wider public remains overall ill-informed and politically apathetic, or at best contributes to the arena on a purely emotional basis. In fact, scientists seldom agree on the causes, effects and future of environmental problems. This lack of agreement reinforces the belief of the public and the decision makers, that scientific opinions are unreliable.

To prevent this unproductive situation, scientists in the highly socially relevant and significant field of global environment should reflect further on their role in the decision-making process.

This chapter is an attempt at such a contribution.

The State of Knowledge on the Global Environment

It is important to realize how much scientific knowledge on the present and future state of the global environment is now available and what the degree is of its certainty.

The fundamental question is whether gases released in the atmosphere will cause global climate changes which will threaten the biosphere and human civilization. This question appears to be full of scientific uncertainties.

The cognitive foundation for the concern on global environment stems mainly from the meteorological theory of the greenhouse effect and on the empirical data on the amount of greenhouse gases mankind is adding to the atmosphere. The greenhouse theory explains the relative equilibrium of incoming radiation from the sun, and outgoing radiation which is reduced through absorption by trace gases in the atmosphere. Higher levels of these gases will theoretically lead to higher temperatures. It is well known which gases have this effect and approximately how much of them man is adding to the atmosphere. It is also known that we produce these gases mainly by the burning of fossil fuels. Other things remaining constant, such as for example present policies, global warming can theoretically be expected. For many, this theory and these facts are sufficient to feed the growing concern on our global climate and to justify countermeasures.

The theory however, sound as it may be, is too abstract to describe the complex systems of global climate. It doesn't include all relevant influential variables and more specifically it doesn't include knowledge of the sinks of the produced greenhouse gases and their spatial distribution.

Many questions remain open concerning the nature of the problem itself. The factual description of the global problem is still in its infancy. The available data on carbon dioxide in the atmosphere are obtained from a geographically limited number of sources and they are measured over a historical period which is too short to justify a reliable long term prediction. No certainty exists about the anthropogenic ratio of the causes of global warming. The natural flow of carbon dioxide is still 20 times higher than the anthropogenic portion.

The degree of actual temperature rise is also uncertain. The global mean surface temperature rise is around $0.5^\circ K$, which falls within the order of measurement errors over the measured period. It is still difficult to separate the real temperature rise from long and short natural cycles and noise. There are great differences for various locations and atmospheric layers.

The theory based climate models do not yet incorporate many uncertain but important factors, such as the behavior of the oceans, the influence of clouds, the cooling effects of aerosols and the growth of vegetation in colder climatic regions. Important positive and negative feedback mechanisms are not yet identified and quantified.

The empirical validity of the models is based on their reconstruction of historical climate data. This can be expected since they are constructed partly on these data. This, however, doesn't guarantee their validity for future changes in climate, if the underlying systems are non-linear and unstable. Predictions with these models are necessarily conditional on inputs such as emissions of gases. This uncertainty is sometimes expressed in scenarios, defined by these inputs.

Given these weak climatological premises and the lack of methodologies for the assessment of climatic impacts, high confidence is not justified in statements on the meteorological and biological effects of possible climatic changes. (Such predictions are often made concerning productivity in agriculture, marine life, latitudinal shifts in vegetation, rise in sea levels, storms and droughts.) This is particularly so where time lags between causes and effects are large and the distribution of the effects over regions may differ widely. Given these uncertainties about the social and economic effects, one can do no more than speculate.

Apart from difficulties in the determination of the effects, their evaluation is very dependent on the value judgments of the parties involved. Cost benefit analyses cannot be calculated for specific regions over longer terms.

These uncertainties make it difficult from a scientific point of view not only to determine what measures and actions should be taken to prevent or cure global environmental problems, but also how these measures should be implemented and enforced.

A final open question is how, given these scientific uncertainties, political support for such measures can be gained, especially if one considers the diversity of impacts on regions and of interest groups.

Suggested Solutions

The solutions for the environmental problems brought forward by scientists are of a different nature:

On the most fundamental level, the problems of global environment are the product of a growing world population and per capita increasing production and consumption. The problems are increased by production methods, which are still very inefficient, in terms of the amount of materials and energy used per unit product, and polluting, in terms of the nature of the waste products. The inefficiency is often explained by the absence of efficient technologies or by their under-utilization as result of imperfectly working markets. The latter can be due to a lack of information, interventions from government or powerful interest groups or central industrial planning by communist states. The effects are aggravated because the capacity of nature to absorb and neutralize the waste is affected by man's actions, for instance in the case of carbon dioxide by deforestation.

In theory solutions to the problem of global warming are indicated by these basic levels:

The growth of the world population is not easy to control as long as people have the potential and the fundamental right to procreate, and as long as for many people this remains the sole means for care in their old age. Moreover in the medium term much effect on global warming is not to be expected.

Slowing down economic growth is politically not a viable solution as long as the present gap in living standards between the majority of the world is population and the rich countries remains, and as long as these rich countries continue to be involved in an economic race for high stakes.

The most promising approach is undoubtedly the development and implementation of technologies for increasing the efficiency of energy production and consumption. The adoption of these kinds of technologies will be promoted to the extent that these technologies are also cost effective. The possibility of a further shift away from fossil fuels in the mix of alternative energy sources deserves more attention and effort. In the short term such efforts will meet considerable resistance from inertia and entrenched interests.

The improvement of energy efficiency or waste reduction by the better working of the marketplace is a partial solution. The market can be useful for the transfer of energy efficient technologies as long as they save money. Its functioning can be improved by providing independent information. But to collective decision makers should be left those actions which individuals in the market cannot take as result of their shortsightedness or their short term interests. Furthermore, important participants such as future generations and nature itself are not parties in the market place.

Another factor is that not all assets of a healthy environment can be expressed in market prices.

The reform of a centrally planned economic system follows its own dynamics and will be more determined by ideological and economic factors than by global environmental considerations.

The Limits of Scientific Research

This list of uncertainties on the problem and its solutions provides no justification for a policy of "wait and see". The greenhouse theory and our production levels show that it is possible for man to alter the global environment substantially. This provides sufficient reason to strive for better understanding of the problem of global warming and its possible solutions, and for implementing adequate counteractions.

It should further be remarked that every scientific endeavor which warns of possible disasters can act to limit them. In that sense the pragmatic value of such research on global environment can easily surpass its cognitive value and should thus be welcomed.

As the philosophy of science tells us, there is no sharp demarcation between science and other types of knowledge. In Poppers' opinion science slowly progresses to higher levels of general understanding by processes of trial and error. Knowledge will always be limited by temporal and spatial restrictions and rationality will be bounded by the limitations of the human brain, logic and computing machines.

This awareness of its limits doesn't have to paralyze scientific research, but could bring it to admit that it will not be able to provide all knowledge of future states of the global environment. This also holds for technology. Engineers cannot pretend to be able to provide the technologies needed to fix all environmental damage. Such forms of hubris also neglect the fact that some of these effects may be irreversible. Further, social and political scientists will never be able to prescribe the political measures which should be taken to protect the environment adequately.

One of the reasons for these limitations is that global environmental problems are multidimensional and complex and require an approach which crosses the traditional boundaries of the scientific disciplines. Their solution requires interdisciplinary and multidisciplinary attack and will not be successful without the participation of all relevant agencies.

The Relation Between Science and Policies for Environmental Protection

The effectiveness of policy measures to protect the global environment depends on more than scientific knowledge. The introductory speech of the conference contains the implicit suggestion that increasing scientific information and prediction will eventually lead to a better control of environmental problems. This idea reflects the well known linear sequence of scientific information, explanation, prediction and control. This model appeared to be very useful in the field of the natural sciences, but may be of lesser value in solving the problems of the global environment.

These problems are ill defined and ill structured. Science is plagued by limited rationality and insufficient data. Models are too simple to capture the reality. There is a lack of general explanations and theories, and predictive powers outside the realm of natural science are very weak. Moreover the powers of decision makers are not strong enough to be able to control global problems. This is demonstrated by two world wars, recurring famines and the arsenal of nuclear arms.

In addition policy-making is not necessarily driven by the simple scheme of a (future) problem-action relationship. Policy makers must consider and weigh complexes of problems, not only of an environmental nature, but also for example of an economic character. Their freedom of action is restricted by socio-political constraints, by other agents, and by supporters or opponents in the political game. They must choose between a limited number of viable options. Understandably they tend to pay more attention to urgent, short term and concrete problems, than to possible future but uncertain problems. For important future problems they agree more readily on verbal intentional statements than on immediate concrete actions.

Decision makers possess specific knowledge not only about the environment and about other related fields, but also about their political support, constraints and priorities. If this knowledge could be made available to scientific researchers they might well perceive the problem from other perspectives, which could result in a more realistic problem definition. This does not mean that scientists could or should replace decision makers. Scientists, by definition, do not share the same responsibilities. A closer relationship between these parties however will promote a more thorough understanding of the problems at hand and of the possible solutions. Decision makers cannot permit themselves to be spiders in a web but must communicate their ideas and knowledge to all groups involved in order to test them out.

Scientists should communicate at an early stage with these groups in order to formulate a more valid problem definition, to give a more relevant direction to their research and to increase the acceptance and utilization of their findings. This does not necessarily mean that scientists should give up their scientific autonomy or integrity. But the position of free floating intellectuals, abstracted from the complex problems of real life will not be very fruitful either.

What is important in the final instance is the translation of knowledge into effective policies and actions. These solutions are completely dependent in my opinion on the close relationship between decision makers, scientists and the wider society. In this triangle, scientific researchers in this highly socially relevant and significant field of the global environment should pay greater attention to their role in society and their contribution to the decision-making process.

A Japanese Example

The Japanese method of pollution control is often promoted as an example of good environmental management.

After several environmental disasters, demands for environmental protection were made by pollution victims, concerned Japanese people and environmentalists, supported by scientists, the press and the judiciary. Successful litigation spurred private industry to support legislation. This led eventually to strict environmental regulations and provisions for their enforcement. The Basic Law for

Environmental Pollution Control aims, apart from economic considerations, to protect the health of the nation and the living environment. The Environment Agency, as a part of the Prime Minister´s Office, has the task of formulating and promoting basic principles for the conservation of the environment and of coordinating other relevant Agencies. The control and power of enforcement is in the hands of relevant ministries. The instruments are health damage compensation, antipollution agreements and environmental monitoring and reporting. As a result of political, judicial and moral pressure industry is developing pollution-preventing technologies and following antipollution agreements and regulations. The Central Council for Environmental Pollution Control is charged with advising and directly reporting to the Prime Minister on environmental matters. The Council is composed of scientists, former bureaucrats, representatives from local government and private enterprise, labor unions, the press and other community organizations. Moreover there are three research institutes attached. This structure forms in my opinion a good example of cooperation between scientists, decision makers and the public.

The problem of the changing global environment can be taken as a problem of global security. Politically this could be expressed as a need for international structures with the power to intervene in the domestic policies of member states, if they endanger the global human environment. The Japanese example could well inspire thinking about international structures for integrating science and policy-making on global environment issues.

Panel Discussion

Panel Chairman: Y. Omoto
Panel Members: Y. Fukuoka, T. Schelling, S. Tsunogai, P. van der Staal,
     C. Lincoln
Panel Summary: H. Krupp

Format:

At the beginning of the morning each panelist gave a short paper focusing on the central theme "What should be done?". A reduced version of these papers is given here. The papers were followed by general discussion which was summarized at the end of the session by Prof. Krupp.
 During the morning proceedings were interrupted to allow Mr. Kazuo Kato of the Environmental Agency to announce a major decision of the Japanese Government on global warming. A summary of this announcement is given at the end of this section.

Panel Discussion

I am Omoto from the College of Agriculture of the University of Osaka Prefecture.
 Now, this afternoon´s session: first, each of six speakers will speak for 10 minutes on his subject, keeping in mind that the central theme of this discussion is "what shall be done".
 After the 10-minute speeches, which will take altogether one hour, we will have inter-panel discussions. The main theme will not be decided, but I hope their talk will lead to a final decision for the chairmen.
 The first speaker is Prof. Fukuoka, professor of Climatology and Air Pollution, Department of Environmental Sciences, Hiroshima University.

Introductory Papers

Dr. Y. Fukuoka:
During the latest decade, I have studied environmental problems in cities. And some of these problems are connected with air pollution.
 I would like to speak today about two topics: the first is the "heat island" phenomenon as a local-scale warming; and the second is acid rain caused by some of the greenhouse effect gases.
 This spring, we began to observe routinely the air temperatures 20-30 m above ground level using automatic recording thermometers at 24 stations, at 2 p.m., and in the early morning. At both of these times we found one "heat island" at roof-top level. Near the ground many small heat islands were observed, as if the one big heat island above, is divided at ground level into small heat islands by rivers, or

The Global Environment
Editors: K. Takeuchi · M. Yoshino © Springer-Verlag Berlin, Heidelberg 1991

the green belt zone. It seemed that rivers or green belt
zones had a cooling effect on the air temperature.

Now for acid rain. From January 1st of this year, I have
measured the pH value of rain water for every single rainfall
in Hiroshima City. It is noteworthy that lower pH values -
below 4.4 - correspond to yellow-sun-weather days. On yellow-
sun-weather days, the sky turns yellow in some areas in Japan,
especially from the end of winter to spring. We think the
weather charts show that this yellowing of the sun in
Hiroshima City is caused by the "borderless transportation" of
acid pollutants from the Chinese continent. We think that
most Chinese factories and oil plants have no facilities to
remove sulfur and nitrogen. Of course, we have many domestic
sources of acid pollutants in Japan also.

We can conclude that acid rain damages green zones in
cities, and that decreasing the green zones will intensify the
heat islands or heat pollution. I have read that the origin
of the word "green" is the same as that of "grow", and I think
that without green, you cannot expect the growth of a country
or nation. In conclusion it is my belief that we human beings
should be kind to Nature.

Dr. T. Schelling:

Ordinarily, I would expect that, in deciding what to do about
carbon emissions and global warming, one would attempt to
assess the damage that may be done and assess the cost of
averting it. It is interesting that the cost of reducing
carbon emissions has been the subject of careful study in the
United States by a number of economic institutes since the
middle or late 1970s. But almost nothing was done that I am
aware of, outside of the field of agriculture, in trying to
estimate the impact of climate change on real income or the
standard of living. Only in the past two years has any work
been done that I am aware of, and most of that has been done
in and on the United States.

And what it looks like is that it doesn't make any
difference, apart from agriculture, fisheries and forestry.
This result comes out of the simple observation that, in a
country like the United States - very little economic activity
is affected at all by the weather. Not manufacturing; not
mining; not electric power production; not banking; not
medical care; not education. Transportation and construction
can be a little bit affected, but on the whole, if the effect
is warming, they will probably be benefitted.

The serious impacts, if any, appear to be in agriculture,
fisheries, and forestry, which amount to less than 3% of GNP
in the United States. I think the latest estimates are that
the impact on agriculture is as likely to be favorable as
unfavorable. I observe that, as in Europe, in the United
States the agricultural problem for 50 years has been to get
rid of surpluses. It's extremely unlikely that anything could
happen to agriculture as drastic as a one-third reduction in
productivity. But in the United States a one-third reduction
in productivity would be 1% subtracted from the GNP. And if
that 1% were subtracted from the GNP in the year 2050, then
instead of being, say, 250% of what it is now, it might be
245% - a difference that you couldn't notice.

It's very difficult to see any serious impact on health and
comfort. I observe in the United States that, for the last 50

years, the general direction of migration has always been from colder to warmer parts of the country.

I don't see any reason why the results we've come up with should be very different for Western Europe and Japan. And this leads me to a conclusion that the only countries that can afford to do anything about carbon dioxide may not find that they have any material interest of their own in doing so.

What about the rest of the world? I think that countries like China and India and most of the countries of Africa - all of the developing world - have so much at stake in the need for economic growth and improvement, that to invest anything, or to sacrifice anything, in the hope that the climate will be more benign 50 or 75 years from now, would probably be short-sighted on their part. The best thing they can do to make themselves less vulnerable to climate change in 50 or 75 years is to stop being so poor.

What about Eastern Europe and the Soviet Union? Well, there, I think the entire world has such a large stake in the economic success of those nations, that for them to make any kind of economic sacrifice, merely in the interest of holding down carbon emissions, would be a mistake from their point of view and from our point of view. And if they did have a margin of resources to put into environmental clean-up, the urgent need is to clean up all of the toxins and metals they've been putting into their water and soil.

This leads me, in a way, to the optimistic conclusion that it may not matter much; to the pessimistic conclusion that, in the case it does matter a lot, I think it's going to be a long time before anything is done.

Let me talk a little about what it might cost. The estimates are that, for the United States, to hold emissions constant at about the level of the year 2000, would cost between 1 to 2% of GNP in perpetuity. I think it's perfectly clear that, if the entire world were to try to stabilize emissions, they will be stabilized in developing countries only at the expense of North America, Japan, and Western Europe. That might mean it would cost us not 1 to 2%, but 2 to 4%.

And now, one of the questions is, "Is 1 or 2 or 3 or 4% of GNP a big number or a little number?". Well, if you think that GNP will grow at 2% per year or more, then to lose 2% of GNP means that, in the year 2050, we will have only the GNP that we were supposed to have in 2049.

I think it's the case that it's the developing countries that are indeed terribly dependent on agriculture; terribly dependent on food production that would benefit most. They're more vulnerable because they're more dependent.

The final question is: If we were prepared to spend 100 to 200 billion dollars per year to make developing countries less vulnerable to economic disaster in the middle of the next century, would we be wiser to invest it in carbon abatement, or to invest it in education, public health, infrastructure, reduced pollution of other kinds, directly in the developing countries? I have no doubt which they would prefer. They would hope to be much less vulnerable to climate change if they could speed up economic development now.

Dr. S. Tsunogai:

Yesterday, Prof. Stewart pointed out the important role of the ocean in climate change, because the ocean is an extremely large reservoir. The ocean contains 50 times as much carbon dioxide as the atmosphere, if only 2% of the carbon dioxide in the ocean enters the atmosphere, the concentration of carbon dioxide in the atmosphere is doubled. I can't say that this situation is an impossibility in the near future, because the ocean is not merely a store, but a huge chemical factory which produces such substances as organic carbon and calcium carbonate. But most important of all is the feedback process. Feedback processes in the carbon cycle in the ocean are not included in the present climate change models. (No current models account for them.) Evidence of the feedback process is found in the record of the ancient earth.

A study published recently in "Nature" reported on past global change recorded in ice sheet in Antarctica. As is well known, the earth repeated glacial and interglacial cycles periodically, and about 120,000 years ago it was warm. During the warm period, the concentrations of atmospheric methane and carbon dioxide were also high. The initiator of the high methane and carbon dioxide levels was not man's activity but solar irradiation. However the level of solar irradiation changed only slightly and this change alone would have been insufficient to induce the temperature increase. A small change would induce a small increase in the concentration of greenhouse gases in the atmosphere, which would then increase the air temperature a bit more. This cycle would have been continuously repeated and amplified. The increase in the greenhouse gases was thus a result of the feedback process. I am convinced that the cause of global warming lies in the ocean, because the ocean is a fluid and contains much carbon dioxide.

To identify the feedback processes is the same thing as finding the fate of the missing carbon. Every year 6 gigatons of carbon dioxide is released into the atmosphere, and 3 gigatons of carbon dioxide remains there. The other 3 gigatons go missing and must be absorbed either into the oceans or into the land. I do not believe the land is a sink for carbon dioxide, but rather a source, because of deforestation and the oxidation of soil carbon.

The deep water formed in the North Atlantic carries only about one gigaton of carbon dioxide caused by human activity. We call the other two gigatons the "missing carbon". I believe that the missing carbon is absorbed into the "intermediate" waters, and particles of organic carbon are produced in the coastal and hemipelagic seas by biological activity. However this idea is not accepted by most scientists. Today I would like to explain my thoughts.

The low-salinity intermediate water flows from south to north in the southern Atlantic, Pacific and Indian Oceans, and from north to south in the North Pacific. The intermediate water, which has a life time of a few decades, is produced in winter by the sinking of cold water, when the air-sea exchange of gases is extremely active. These exchanges are highly dependent on the condition of the sea surface and restricted to times of severe storms. The partial pressure of carbon dioxide in the winter ocean water is low, because surface water temperature is low and the carbon dioxide has been consumed by organisms during the summer. The intermediate water carrying much carbon dioxide develops over a fairly vast area. Organic carbon fluxes in the ocean are not formed uniformly: fluxes in high latitudes are one order of magnitude larger than those in lower latitudes. These results were obtained from my research work with sediment traps. In coastal seas the flux is much larger still. Organic carbon particles are produced exclusively in coastal sea and in the subarctic oceans. They decompose in the bottom of these shallower seas, and in deep water containing fine particles of organic matter. The water carrying increased carbon dioxide joins the intermediate water and the deep water.

There are other serious problems which concern the ocean, but I don't have time to discuss them today. Finally, I would like to say that biological activity does not necessarily act as a sink for atmospheric carbon. Contrary to the formation of organic carbon, the formation of calcium carbonate acts as a source of atmospheric carbon. In the ocean, organic carbon is decomposed quickly, but calcium carbonate is preserved at the bottom, especially in shallow seas. The decomposition products (nutrients), are again used for calcium carbonate production. Although eutrophication without silicate due to human activities increases the rate of formation of organic carbon, that of calcium carbonate increases much more. Thus changes in the ocean may accelerate the increase in atmospheric carbon dioxide.

Since the life time of the intermediate water is about 30 to 50 years, the two gigatons (if my idea is correct), of carbon dioxide per year now contained in the intermediate water, may be emitted to the atmosphere after a few decades. I conclude that an increase in atmospheric carbon could be accelerated in the not too distant future by this effect.

Dr. P. van der Staal:

I have heard many interesting speeches these past two days, and I think that there is a lot of scientific knowledge already available for making good decisions. But something I have gathered is that there is a belief that if you have enough information and enough knowledge, you can make good predictions, and then automatically you can control the phenomena which are causing your problems. But I think that in social problem fields, the situation is more complex, and that we should try to find additional means to deal with them.

I think that the effective solution of environmental problems is not only promoted by trying to find new technology, but depends on better interlinkage between the groups involved. These groups are: scientists and knowledge generators, decision makers, and the general public. In my opinion, scientists are sometimes not well understood, and scientific knowledge is not very well used by decision makers. Sometimes scientists send their information mixed with contradictions to the decision makers, and they feel frustrated if their ideas are not used.

I believe that scientists should reflect on their role in society. We need better communication and cooperation with decision makers; we need a well-educated public which is committed and conscious; and we need the participation of politically-unconstrained media, if we are to achieve good political measures. I believe that in Japan there is a

tradition of a better cooperation between scientists and
decision makers, and that the public also plays an important
role. For instance, public opinion concerning science and
technology is asked for in opinion surveys.

So what I want to propose is that we should find ways for
scientists to participate more in decision-making processes,
in order firstly to understand better what the specific
problems of decision makers are, and secondly to provide
decision makers with better scientific understanding, so that
scientific advice can be translated into political measures
which can hopefully be enforced in order to solve
environmental problems.

Dr. C. Lincoln:

I just thought I would show something that is pretty obvious:
that whatever we do in regard to the environment, it's got to
be an independent action which is based on participation and
crosses all boundaries.

First of all the individual is always at the center of
everything. The figures for light and heating of people's
homes in the United States is 77 billion dollars, so you can
see the individual has got a huge part to play. But nothing
can be done by the individual alone; it will have to be a
shared thing with the community at large.

Everything is interlinked, but certain areas strike me as
being particularly important: science and technology, energy
and transportation, social sciences, health, and of course
arts and culture, and finance and economy. And I reserve a
special place for education, which leads to awareness. And of
course, information and consultation with the public. And
then planning leading to decisions. And throughout it all,
ethics. I feel that if we start from the bases of ethics and
responsibility, and the right to a decent quality of life, we
will put the environmental problem in focus, both on a local
and global scale.

I'd like to mention some things which relate to my own
experiences in government and political life, where
traditional concepts of vertical government and decision
making no longer work. All these deals of vertical lines
where each person in charge of a department or ministry pushes
his own little empire, without any interaction very often,
are, I think, one of the very key reasons for the problems we
know today. I suggest that successful, sustainable government
action depends on close interaction among departments and
decision makers, and on enlisting the involvement of key
sectors of activity. As a concrete suggestion, I could
mention that we have been trying in Canada: round tables.

I have a few other suggestions before I close. I really
believe that one of the sad things is that information is
always available somewhere. We just don't know where it is.
And even those of us that are part of this network sometimes
cannot find it. I think there must be some form of
international network - maybe it should be called UNET. Some
way of polling, gathering and distributing information on
research, on studies, on emergencies especially, and emergency
procedures.

Another suggestion is that we need a means of carrying out
objective expert analyses, risk assessments,

mediation/arbitration.   I'm talking here of panels - of
voluntary panels - of expert individuals.

As a third suggestion, I think one of the key and critical
environmental challenges ahead of us is the urban environment,
as the whole population of the world moves toward cities,
which become city-states.   If we had an urban planning model
for the world, including mass transit, environmental sectors,
energy looming large, of course, water, sewage, domestic
waste, and so on.   Wouldn't it be great if we could build that
model?   Maybe the Japanese with all their resources and
inventiveness and creativeness could do it.   I think it's a
very, very important issue.

Finally, I suggest an international conservation fund, the
basic funding of which would be generated by every developed
country to the extent of 1% of its present defense budget.
Total world defense spending today is one trillion dollars,
and just 1% only of this budget could provide 10 billion
dollars in a conservation fund.

As a last suggestion, I think that this meeting should
express a strong conclusion on the urgency for energy-wasting
countries to adopt conservation measures.   Countries with a
means to do so should accelerate alternative energy use as a
replacement to fossil fuels.   These are the suggestions I want
to bring forward.

General Discussion

Dr. T. Schelling:

"I think that what we have to now is begin looking at goals,
and therefore actions, or steps, to try to meet these goals.

The first goal I think is the need to conserve energy.
This will do a lot of things, will have a lot of benefits, and
there is absolutely no excuse to be throwing away the
resources that we are doing now.

A second goal has to be reduce emissions - atmospheric as
well as liquid and other toxic wastes.   We should reduce
carbon dioxide emissions; reduce SO_x s, NO_x s, etc., as well as
toxic wastes, and of course the CFCs.

Another thing we have to try to promote is stopping
deforestation, and encouraging reforestation.

A final goal, I think, should be to do something about
population growth.   That would mean, in the first instance, at
least reducing population growth.   Now, Professor Hussein very
kindly showed us that graph where not only we have the
populations of the various regions, but we have the energy-
equivalent population.   I think that's a very powerful
message.

Now how could we try to reach these goals? One way would
be, of course, research. We have to know more about what we
are doing; about the world as it is; and about things we can
do to change.   A second step is to find ways to re-orient
policies of government, of decision makers, and of companies
as well. We must also transfer technology to those who don't
have the means to pay for it.   There may be more that you will
want to include, but this is what I propose."

Dr. P. van der Staal:

"There are two examples of experience we have had in Holland of a bad relationship between scientists and decision makers. We have had a wide societal debate on nuclear energy in Holland where the public was invited to give their opinions about how the government should proceed. The public in general was of the opinion that we should stop nuclear energy production, but the government ignored the voice of the Dutch people. The politicians said that the public was not educated well enough in scientific and technological aspects to give an opinion which should be honored by the decision makers. So what I think is that the principal politicians have some right to ignore scientific studies and public opinion, because the public should be educated. And I think scientists have a specific role in educating the public. I think scientists should be more aware of the complex problems of decision makers, who have a lot of problems, sometimes contradictory to weigh up. There should be better cooperation between these groups."

Dr. Takeuchi:

"I think there are two questions which are not value judgements which we have to clarify. The first is whether global warming could prove to be serious or not. Of course, there are some uncertainties, but I think that there is some probability that something serious could happen if the present trend of CO_2 emissions or the emissions of other gases continue to grow at the present rate.
    The second question is whether there are any feasible technological means by which we can stop increasing emissions, without decreasing our economic welfare or stopping economic growth. My personal answer to both of these statements is yes. But it seems that Dr. Schelling is against both of these propositions. I think he is going to say that we will lose a lot of money by simply worrying too much about something which may not be a real danger. Also Dr. Schelling is implying that there is no possibility of emission control without very, very high cost, which I think is very contrary to the statement made by Dr. Lovins. Of course it is not necessary to have unanimous views on any of the statements, but I think that we should first clarify what we think about these two basic questions.
    Also, although in this room, maybe with the exception of Dr. Schelling, most of us are going to say "Yes" to both of the questions, for others outside this room that is not necessarily true. I think the American government´s attitude has recently switched to being strongly "No", that the global warming issue is not so serious."

Dr. C. Lincoln:

"Let´s concentrate on ideas where we can find a meeting ground, rather than spend 20 minutes to find out whether we agree or we don´t agree whether CO_2 is causing global warming."

Prof. Hussain:

"I want to concentrate on public policy-type problems. I would like to suggest that if global warming is considered an immediate problem, population control is not a way of solving that problem, although there may be good reasons for controlling population for other reasons. Secondly, too drastic a decrease in population growth has catastrophic social consequences, because it makes the age structure of a population highly adverse. So I would like to suggest that, in trying to solve one problem of global warming, we should not fall in the trap of neglecting everything else. We shouldn't come out of this room, having agreed to population growth as a way of solving the problem of global crisis, when in fact, we can see quite good arguments that it certainly doesn't provide any immediate answer to the problem.

The second thing I would suggest is that quantitative targets do not make policy. There is quite a lot of work and research which needs to be done to translate these quantitative targets into, for example, the kind of price changes or tax changes which will be required.

The third thing is that before we agree to having set a world fund or something similar, we need to take into account what will happen to previous international agreements. The record of the world community is actually not all that good in adhering to agreements it has arrived at in past."

Dr. Stewart:

"I guess maybe it's a Canadian pragmatism which comes out here, and I am in agreement with Mr. Lincoln. I think we really should look at what we hold in common, and not concern ourselves greatly about where differences may arise. The view which I have been advocating for some time is that we should look among the actions which we can take, and choose those options which make sense, those which have least probable effect on the future of mankind. I should say, by the way, that the reason we are so pragmatic in Canada is we can't afford to be idealistic and logical, because we live in an illogical country!"

Dr. Bach:

"I think we need a kind of strategy that solves many problems simultaneously. One can even make a priority list of possible objectives. In fact that list exists. Of course it comes from the Enquete Commission; which is where I spend most of my time. In the first place (there was no disagreement whatsoever) comes more efficient energy use. By achieving that, you solve almost all of the problems you have around. This is important, in industrialized countries and it's even more important in the lesser developed countries.

Secondly comes the reduction of CFCs. This is pretty much advanced already. It is not licked yet, because the money has not been appropriated and without that, China and India will not go along, and these are key countries in the game. There are relatively few actors that produce this problem. By reducing CFCs you solve several problems - not two - not only the ozone hole, and not only the greenhouse effect. There are lots related to that. If the protective ozone layer is

reduced and more ultraviolet light gets through, then you have
a lot of trouble with human health with phytoplankton in the
ocean, with the growth of green plants on the earth. So one
solves a great number of problems by this option. So it
should have a high priority in our listing.

Thirdly comes the reduction of fossil fuel use, for many
reasons, including ethical ones. Our descendants will curse
us that we have used up these important non-renewable
resources, and not left them for the thing that they should be
used for, namely, the presently 4 million different things we
can make out of carbon. Carbon should be reduced, for many
reasons, including for the reason of reducing CO_2. At the
same time one reduces the SO_x, the NO_x, and the CH_4, and the
CO - what they call in Europe the "summer smog" - the acid
rain.

Next comes the reduction of destruction, not only of
tropical forests, bus also of forests everywhere in the world.
This has so many implications, particularly ethical ones
again.

Next in line comes the reduction of the emissions that
particularly pertain to agricultural usage, and they are CH_4
and N_2O - very important greenhouse gases. N_2O is a very
important gas as far as the destruction of the ozone layer is
concerned. And, of course, the oversupply of artificial
fertilizers destroys not only the soil - it also destroys the
water.

And the last step is to develop sources of sustainable
energy. We have neglected this for so long.

So these are the kind of very interrelated problems, on
which I think we can achieve an agreement."

A voice from the audience:

"Just one question for Dr. Schelling. I've heard quite a lot
now of public polls into how people perceive these sort of
questions. Would you agree that there is a changing climate
for politics here that we couldn't neglect?"

Dr. T. Schelling:

"I'm not much of an expert on public opinion, especially
outside the United States. I think there is a great deal of
popular enthusiasm for doing something about the greenhouse
effect in Western Europe and in a lot of other countries. In
my own country, I think the likely response from the public to
any call to do something about greenhouse warming we've
probably already seen: an absolute rejection of a 9-cent per
gallon tax on gasoline. If we wanted to be serious in the
United States, we would at least propose that there be a
federal tax on gasoline that increased by 10 cents per year
for 10 years.

So my worry is: I don't think much will be done by Western
Europe and Japan if the United States doesn't go along. But I
don't perceive, in my country, any great environmental
interest in doing anything that hurts.

Now, if I could just take one more minute to clarify my
position. Firstly, I did not mean to belittle in the
slightest the likelihood of climate change. What I was saying
was that the only analyses I've seen have to do with
agriculture. I invite anybody to demonstrate what the impact

on standards of living would be. Now, I also thought I was
saying that to me it looked cheap to abate carbon emissions.
The question is, why should one want to do this and I think
there are a couple of good reasons. One is that there may be
some very important fragile ecosystems that we would like to
preserve - biological diversity and things of that sort. I
think a more important one is that I don't trust the general
circulation models, and the reason I don't trust them is the
people responsible for them have advised me not to trust them.
And therefore, I would think if the cost of insurance is to
postpone getting to where we wanted to be in 2050 until 2051,
it is a trivial price to pay. What I find very implausible is
that my countrymen will be willing to take that view until Dr.
Krupp's "generation time" has passed. I don't think the
United States is serious."

Prof. Douguedroit:

"My intervention concerns public opinion. Because public
opinion in many countries has a large influence, for example,
on the level of consumption of fuels. But I think that in
many countries public opinion is not well informed about the
effects of gases - gas emissions, CO_2, methane and so on. And
public opinion has a very important influence on decision
makers, especially politicians. So I think that we should
propose the wider dissemination of information, independently
of decision makers. And I strongly believe that the schools
should be targeted first by such a general information
campaign. The problem is for me that the general level of
understanding in most countries is not good enough."

Dr. Kellogg:

"Yes, I would like very much to have a go at my old friend Tom
Schelling here, because in listening to him I heard a great
deal of the kind of argument that our White House uses in
trying to downplay the need for the United States to do
anything about the climate change. The US has resisted any
temptation thrown up by the Western European countries to
start some kind of an agreement on the reduction of fossil
fuels. President Bush himself stood up before the Council of
Ministers and said this.
    Now I would like to just summarize my point of view on
this, Tom. I think the global warming is real, although the
details of the timing and the regional changes are still not
clear. I'm convinced that climate change will surely affect
rainfall patterns, and this will have big regional influences,
not only on agriculture, but on natural systems. Canada, for
example, has enormous forest resources, and they're very
concerned about the fact that if you move the climate zones,
the forest will be left behind in unfavorable climates, and
they will die out. Forests simply can't migrate at the rate
of change that we're talking about.
    So this is just one of the problems that I see. There are
other problems associated with global warming, such as the
rise of sea level - not as much as was claimed earlier, but
nevertheless a large fraction of a meter by the end of the
next century. It will be expensive.

So in summary, I think that the actions that we could take in the US, Japan and the Western European countries - I'm speaking now from the point of view of the industrialized world - those actions seem to make sense. We know that the biggest lobbies in Washington, Tom, are the petroleum and automotive interests. Imagine sitting in the White House with this at your back. You would be afraid to raise gasoline taxes. I think there's a power play here, and actually the people are not that hostile to the idea of raising gasoline taxes - it's the politicians."

Dr. Lovins:

"This is for Tom Schelling. I would be the first to agree that a dollar a gallon extra gasoline tax is probably a good instrument of externalities, and I would be the last to quarrel with economic glasnost, that is, "prices that tell the truth". However, raising the gasoline tax, although it probably is a good idea, is the least effective way to get efficient cars on the road, because only one-sixth the cost of driving at "pre-Sadam Hussein" prices is fuel. I would place a lot of significance in the votes of 61-11 in the California Assembly, and 31-2 in the California Senate, in favor of the "fee-bate" I described: a fee or a rebate on new cars, depending on efficiency, because this is the most car-dependent society in the world, and I think Californians correctly recognize that this revenue-neutral way of transferring wealth from those who impose social costs to those who save social costs is a much more effective and sensible approach than simply raising gasoline taxes and having that price signal diluted 5-to-1 by the other costs of owning and driving a car.

It seems to me there are basically two ways to get efficient cars purchased: one is a fee-bate, especially if you base the rebate part of it on the difference in efficiency between the new car and the old car which you scrap; the other way is standards. They're both legitimate methods. The basic point here is simply I don't see any American hostility to getting cars that work better and cost less."

Dr. Shearer:

"I just wanted to reinforce something that Professor Douguedroit has said about education. I had been away from the United States for 17 years, and I returned and began sampling what's going on there. I have contacts with many families there - and something that really amazed me was that children are very conscious of all these things that are going on - all these discussions. I find them educating their parents, and this is and incredible thing. Now, of course, we're talking about Prof. Krupp's 25 years, or maybe a little less. So something is happening, and maybe in 15 or 20 years from now, when these matters come up again there will be a totally different decision.

But I was particularly amazed by education. I think what Prof. Kellogg is doing - at least what he told us he's doing - going and talking to young people, is very important. It's something we all should be doing."

Professor Krupp´s Summary

"I´ve tried to summarize to the extent possible what has been said. I think the first question is, "Is there man-made global warming?" The answer was almost unanimously "Yes, there is." How serious is it? Well, we don´t know exactly, but we must proceed on the assumption that if present trends continue, it´s probably serious. However when it came to the question of whether to take action or not, different views began to emerge.

Action was further differentiated with conditions. The first was that action should have a neutral effect on GNP. If action were to have a negative effect on the GNP, then some people would have different views. Action which had a positive effect on GNP was also canvassed as a possibility. There was no opposition to the list of ameliorating measures given by my German colleague, Herr Bach. They are the result of a long and intense debate, I assume. If we had the same debate here I would assume we would come up with a similar list. So we are quite sure efficiency is good, and therefore, efficiency of energy is very good.

Forest-saving is certainly a good idea. As for emission reduction - I´ve put a little "How?", because in yesterday´s discussions two or three people said "How to control rice paddy fields and cows?"

Then, renewable energy sources - this sounds good, although, of course, technology assessment will be needed to find out what are the side-effects desirable and undesirable.

Population control - the "pro" is obvious to most people here, but to hear Dr. Hussain´s warning about possible negative consequences was a very useful contribution.

Then, financial instruments. National instruments: CO´ tax, energy tax. International instruments would involve trans-boundary money flow. Some things are in the pipeline. The World Bank was mentioned.

Then there was this collection of ideas which come up whenever scientists sit together: R & D is always a good thing; more R & D is always a good thing, too. Round tables - this was meant to be more than just scientists talking; you meant it in a much deeper sense, so inter-sectoral round tables.

Then, greater dissemination of information to the public. This is certainly a good idea.

So the overall conclusions look obvious. These are all things which, as a first step, ought to happen. What was not said in this afternoon´s discussion was, "How about the longer-term perspective?"

In our discussion we remained on the first increment of the analytical continuation of present trends. But, we left out the question, "If we proceed increment-wise by one or two decades or generations, have we even provided enough of an answer?" Well, maybe that´s a longer-term perspective and finds its place in tomorrow morning´s deliberations. Thank you very much."

Dr. Y. Omoto:

"Thank you very much, Dr. Krupp. Thank you very much also to those who gave talks and opinions. And thank you everyone for your cooperation."

Environmental Announcement

During the morning Mr. Kazuo Kato of the Environmental Agency of Japan made an announcement on the decision of the Japanese Government taken that morning to adopt an Action Program to arrest global warming. Mr. Kato indicated that the program will set two targets:
- to stabilize Japan´s national emissions of carbon dioxide at the 1990 level by the year 2000, and
- to stabilize Japan´s per capita emissions of carbon dioxide at the 1990 level by the year 2000.
This announcement was warmly welcomed by participants.

Conclusions and Recommendations of the Working Groups

On the final afternoon of the symposium, participants divided into five working groups to draw up summary statements of the symposium's deliberations. These statements take the form both of conclusions and recommendations.

Working groups were asked to limit their statements as much as possible.

The topic areas of the five groups were as follows:
Group A: Integrating Scientific Priorities
Group B: Environmental Information Exchange and Dissemination
Group C: Risk Philosophy
Group D: Technology Transfer and Efficiency
Group E: Socio-economic Aspects of Relations Between the Industrialized and Developing Countries

Group A: Integrating Scientific Priorities
Chairman: M. Yoshino

It is urgent to promote fundamental research on the following aspects of global warming:
- the constitution of greenhouse gases in the past; future changes from present levels; impacts of likely changes on the environment,
- the processes affecting the cycling of the bioelements carbon and nitrogen in the ocean,
- the long-range transport of pollutants in the atmosphere,
- the possible influence of global warming upon local climatic phenomena such as urban heat islands and fog.

Group B: Environmental Information Exchange and Dissemination
Chairman: W. Bach

Scientific evidence shows that current and potential environmental threats justify urgent countermeasures.

In order to respond to these threats, the relevant environmental information must be made available to the various actors involved: decision-makers, legislators, political parties, lobbyists, environmental groups and the general public.

Even if this information is adequately disseminated, however, a proper balance of power among the actors is required, before environmental protection will follow.

Legislators are influenced by their party platforms, lobbyists, environmental groups and the general public.

The Global Environment
Editors: K. Takeuchi · M. Yoshino © Springer-Verlag Berlin, Heidelberg 1991

Of these, lobbyists and political parties exert the dominant influence, because environmental groups and the general public often have inadequate information and insufficient resources to be effective.

Such mechanisms as referenda, elections and round tables can provide an avenue of expression for environmental groups and the general public, and in this way help to correct the imbalance of power among the principal actors.

Laws to protect the environment are insufficient without provision for independent groups to monitor compliance and enforcement.

Environmental protection must be safeguarded at the constitutional level, as a basic human right, with recourse to a proper judicial process.

Group C: Risk Philosophy
Chairman: W. Kellogg

In planning for the future, full allowance must be made for the fact that the risks may be great and there are many uncertainties. These uncertainties can work in either direction, and their impacts can be either positive or negative. Among other things they derive from:
- the unreliability of climatic, economic and ecological models,
- the potential for unexpected global environmental changes,
- the unforeseeable results of deforestation, and air, soil and water pollution.

To the extent that measures to abate global warming should be taken anyway, for a wide range of reasons, the uncertainties may not be as significant as they seem.

The greatest sources of uncertainty are unforeseeable developments in technology and changes in world order.

In planning for the future state of the world, the priority in policy making for countries, corporations and individuals should be: KEEP FUTURE OPTIONS OPEN.

Group D: Technology Transfer and Efficiency
Chairman: H. Krupp

Market mechanisms have proved highly effective in promoting technology transfer.

When there is an absence or failure of market forces in particular areas, however, public intervention is necessary, to take account of such factors as social costs and the interests of future generations.

Opportunities exist for the mutually beneficial exchange of resource-conserving technologies. They could involve, for example, the transfer of solar technologies to developing countries, and of energy-efficiency technology to Eastern Europe.

Special attention needs to be paid to these factors when selecting technologies:
- the physical, social and cultural character of the region,
- traditional methodology,
- environmental protection.

Appropriate technologies will be selected only if these requirements are met:
- proper technological assessment,
- the enhancement of technological and scientific capability in developing countries,
- the accessibility of information to the public,
- informative labeling,
- measures which will facilitate the distinction between appropriate and inappropriate choices.

Group E: Socio-economic Aspects of Relations Between the Industrialized and Developing Countries
Chairman: F. Duchin

Research and international meetings on industrialized countries and developing countries relations have in the past taken more of a top-down and quantitative approach than a bottom-up and qualitative one. We recommend a shift in emphasis that builds up from specifics in each case.

There is a need to develop workshop formats which will facilitate the interaction of decision makers and legislators from developing countries with scientists from industrialized countries, to consider topics such as energy alternatives.

Population growth is rightly perceived in industrialized countries as exacerbating both environmental and economic problems. Improving the standard of living and the education of women is the surest way to reduce fertility.

Concluding Remarks

There was general agreement among symposium participants that the global environmental problem is an issue of the utmost importance. Researchers throughout the world have repeatedly demonstrated that the global environment nurtures a sensitive and vulnerable ecosystem. It supports all life on earth and can neither be compensated for not replaced from any other source. It is generally agreed that any damage to the environment, whether on a global or local scale, will inevitably worsen human living conditions, as well as restricting the resources needed for mankind´s survival.

Neither destruction nor risk of destruction to any part of the global environment can be allowed. Whatever measures are necessary and possible to preserve the global environment must be carried out.

Many indications have appeared in recent years that increasing production, distribution and consumption made possible by modern scientific technologies, combined with rapidly growing world population, are creating environmental pressures which are reaching dangerous limits. With even greater stresses in the future, the global environment could seriously deteriorate and fatally affect the life of all humankind. Yet awareness of these problems by governments and the wider public is still largely lacking.

Among the many factors that contribute to global environmental damage, some must be accorded the highest priority. They are: global warming and the associated "greenhouse" gases, the changing ozone layer and in particular the "ozone hole", the phenomenon of acid rain and its impacts, and the complex problem of tropical deforestation.

Especially serious attention must be given to the first of these issues, global warming. Its impacts will be many, but one very predictable consequence is a rise in sea levels which could result in a serious threat to the socio-economic conditions of those countries with extensively populated lowland areas. An increase of around $0.5°C$ in global temperatures has been observed during this century, and further warming will most likely lead to increases of between 1.5 and $4.5°C$ by the year 2050. It is predicted that sea levels will rise between 40 and 120 cm by the middle of next century. It is commonly agreed that the main cause of global warming is the increased level of greenhouse gases, particularly CO_2, and that the main source of these is the burning of fossil fuels.

Participants agreed that there are questions of major importance which require further study. In Monsoon Asia regional scenarios related to monsoon and tropical cyclone activities should be looked at. In terms of the impacts of

The Global Environment
Editors: K. Takeuchi · M. Yoshino © Springer-Verlag Berlin, Heidelberg 1991

global warming on agriculture, matters which should be studied include the capability of agricultural technology to respond to climatic change, the tolerance of individual crops, the impact in specific regions on crop type, crop calendars, and so on.

Fundamental scientific studies should not be neglected either, since there still exist many gaps in our scientific knowledge. Basic studies should deal with topics such as: the origin of the greenhouse gases and man-atmosphere interactions; the cycling of bio-elements in the oceans; climatic, economic and ecological models; transfer processes of solar energy; energy efficiency technologies (including multi-sector economic dynamic models with resource and environment constraints).

Although the many uncertainties about the extent and the effects of future global warming need more intensive study, it is generally agreed that urgent measures are needed to counteract continued and increasing deterioration of the global environment.

The need to conserve energy comes first. It has been shown that there are many available means of stopping energy waste which will also have the effect of saving money, since industry would become less polluting. Reducing CFC´s is another of the most urgently needed countermeasures. Rapid population growth is a further cause of global environmental damage. In 1990 world population has reached 5.3 billion. By the year 2050 it is expected that it will have doubled to between 10 and 11 billion. Damage to the global environment will be even greater if no effective steps are taken to limit the tremendous stress which overpopulation will cause.

Efficiency and wastefulness in energy consumption depend on both socio-economic systems and levels of technological development. We must continue to develop energy-saving, environment-preserving technologies, and we must also improve socio-economic systems so that such technology can be utilized. Global environmental change must therefore be considered from both a physical and socio-economic perspective. In this regard the inter-disciplinary co-operation of scientists is of the greatest importance.

If scientific findings are to be implemented there must be close co-operation between decision- and policy-makers, and scientists. The open unbiased exchange of views must be increased through continuing international conferences. Governmental and non-governmental organizations must also take initiatives. The general public must become well informed about the urgency of taking necessary environmental measures. All nations regardless of their political or economic status, and regardless of their level of development, must work together to accept responsibility for the preservation of the global environment.

Industrialized countries, which have been the main contributors to environmental damage, must accept the claims made on them by the less developed countries for financial assistance with environmental preservation. The developed countries must demonstrate their willingness to take effective measures to control and reduce the pollutants generated within their countries, and to apply energy-saving technologies. At the same time solutions proposed to the less developed countries must not create inequitable burdens on development goals. It should be understood that sustained growth for the less developed countries is not necessarily incompatible with environmental preservation. There exist huge possibilities for technological improvement for such countries, through international co-operation.

Finally, symposium participants agreed unanimously that international and interdisciplinary meetings such as this one are extremely valuable. This conference should be followed by further meetings of a similar nature so that the discussions undertaken here can be continued and extended.

Index of Contributors